Finding Consciousness

OXFORD SERIES IN NEUROSCIENCE, LAW, AND PHILOSOPHY

Series Editors

Lynn Nadel, Frederick Schauer, and Walter P. Sinnott-Armstrong

Conscious Will and Responsibility
Edited by Walter P. Sinnott-Armstrong and Lynn Nadel

Memory and Law
Edited by Lynn Nadel and Walter P. Sinnott-Armstrong

Neuroscience and Legal Responsibility
Edited by Nicole A. Vincent

Handbook on Psychopathy and Law
Edited by Kent A. Kiehl and Walter P. Sinnott-Armstrong

A Primer on Criminal Law and Neuroscience
Edited by Stephen J. Morse and Adina L. Roskies

Bioprediction, Biomarkers, and Bad Behavior
Edited by Ilina Singh, Walter P. Sinnott-Armstrong, and Julian Savulescu

Finding Consciousness: The Neuroscience, Ethics, and Law of Severe Brain Damage
Edited by Walter Sinnott-Armstrong

The Neuroethics of Biomarkers: What the Development of Bioprediction Means for Moral Responsibility, Justice, and the Nature of Mental Disorder
By Matthew L. Baum

Finding Consciousness

The Neuroscience, Ethics, and Law of Severe Brain Damage

EDITED BY

WALTER SINNOTT-ARMSTRONG

OXFORD
UNIVERSITY PRESS

OXFORD
UNIVERSITY PRESS

Oxford University Press is a department of the University of Oxford. It furthers the University's objective of excellence in research, scholarship, and education by publishing worldwide. Oxford is a registered trade mark of Oxford University Press in the UK and certain other countries.

Published in the United States of America by Oxford University Press
198 Madison Avenue, New York, NY 10016, United States of America.

First Edition published in 2016

Library of Congress Cataloging-in-Publication Data
Finding consciousness: the neuroscience, ethics, and law of severe brain damage /
edited by Walter Sinnott-Armstrong.
p. ; cm. — (Oxford series in neuroscience, law, and philosophy)
Includes bibliographical references and index.
ISBN 978-0-19-028030-7 (alk. paper)
I. Sinnott-Armstrong, Walter, 1955– , editor. II. Series: Series in neuroscience,
law, and philosophy.
[DNLM: 1. Consciousness Disorders. 2. Brain Damage, Chronic. 3. Neurosciences—
ethics. 4. Neurosciences—legislation & jurisprudence.
WL 341]
RC387.5
617.4′81044—dc23
2015028971

9 8 7 6 5 4 3 2 1

Printed by Sheridan, USA

PREFACE

As neuroscience gives us new abilities to do what we never dreamed possible, it also creates profound challenges in ethics and law. When we *cannot* do something, there is little or no point in asking whether we *should* do it. But when neuroscience (or any other science) makes it possible to do something novel, then we need to ask whether, when, why, and how we should exercise this ability.

One recent example of this trend is the ability to detect consciousness in patients who have suffered severe brain damage and show no outward sign of consciousness. In the past, families and doctors never imagined that they could communicate with these patients. Then, in 2006, a groundbreaking paper showed that an original method could be used to detect consciousness in patients who had previously been thought unconscious and unreachable. A follow-up paper in 2010 showed that such patients could even answer questions. One patient who had shown no outward sign of consciousness for 5 years answered five autobiographical questions correctly by thinking of motor imagery for "Yes" and spatial imagery for "No." The answers could be detected by brain scanning, although it was impossible to communicate with the patient in any other way. These findings surprised and confused many readers.

The first confusion concerns the precise condition of these patients. We need to distinguish persistent and permanent vegetative states from death, coma, minimally conscious states, and locked-in syndrome. We also need to ask whether consciousness has been or is likely to be found in patients whose current condition is a result of anoxia, brain disease, or traumatic brain injury. The proper diagnosis of these conditions is a complex and uncertain medical issue.

These technical distinctions are crucial for avoiding overgeneralizations such as the conclusion that we should never give up on any patient, regardless of what condition they are in, because every one still might be conscious. That mistake would stand in the way of organ donation and all of its benefits. It would also lead to immense expenditures of resources on patients in hopeless conditions. Of course, we do need to think very carefully about which laws and policies should govern the treatment of all of these patients, but that does not mean that we should treat them all the same regardless of whether they show any signs of consciousness. How we ought to treat various patients is a pressing issue in ethics, law, and policy, and there is no single, simple answer.

The fundamental philosophical question here is, What is it that gives people value and rights? Is consciousness what really matters? Some ethicists have argued that life has no moral status without consciousness, whereas others claim that membership in the human species is enough to confer value, or that pain is bad even when one is not conscious of it. These positions have radically different implications for how we ought to treat patients in whom we still find no signs of consciousness despite our best efforts.

Another complication is that every test of consciousness is imperfect. They all have the potential for false-positive or false-negative findings (cases, respectively, in which the test indicates consciousness where there is none or lack of consciousness where some exists). In the face of such uncertainties, we need to ask how likely these kinds of mistakes are, and which types of mistakes are the worst.

These uncertainties are difficult especially for caregivers, friends, and families of individuals with brain damage. What should they do when they do not know whether a loved one is conscious? And when neural methods detect consciousness, more personal questions are raised: What can caregivers do to improve these patients' lives? What is in their loved one's best interest?

Another issue is whether we should let these patients decide for themselves. If patients can answer questions, would it be more humane and respectful to ask them what they want to be done? This question is different from asking what is in their best interest, because the patient might prefer a course of action that we think is not in their interest. For example, we might think that a patient is better off alive even if severely disabled, but the patient might indicate that he or she does not want to be kept alive. Or the reverse: We might think that a patient's condition is so intolerable or meaningless that he or she would be better off if allowed to die, but the patient might express a preference to live. What should we do in such cases? Are the patient's decisions informed and rational? Are these patients competent?

Such challenging questions arise as neuroscience enables us to detect consciousness in more and more patients with severe brain damage, and all of these issues are discussed in this collection. Most of the prominent contributors met together at Duke University in January 2013 to share their views on these developments and controversies, and then they revised their talks in light of the intense discussions that followed. Three additional essays were added to fill out the picture. The resulting chapters describe the recently developed neural methods of detecting consciousness in patients with brain damage, the contrasts among various conditions in which consciousness is affected by brain damage, the nature of consciousness and its value in determining the moral status of patients, lay attitudes toward letting these patients die, and the many moral, legal, and policy issues raised by these cases. The varied contributors looked at the issues from very different perspectives, informed by different disciplines and methods as well as different ethical and political assumptions. Still, they all agree that innovative methods of consciousness detection raise pressing, important, and fascinating questions about what it is to be human, what is the point of our lives, and which law, policy, and ethical norms should be adopted.

This collection should be of interest not only to academics in the fields of neuroscience, law, ethics, and philosophy but also to anyone with a friend or family member who has suffered brain damage or, indeed, anyone who might suffer brain damage in the future. That includes everyone, because we all have the potential to end up in the conditions that these chapters analyze.

CONTENTS

CONTRIBUTORS

Jeffrey P. Baker
Department of Pediatrics
Duke University
Durham, NC, USA

Tim Bayne
School of Social Sciences
University of Manchester
Manchester, United Kingdom
University of Western Ontario
London, Ontario, Canada

James L. Bernat
Geisel School of Medicine at Dartmouth College
Hanover, NH, USA

Meghan Brayton
Department of Psychology and Neuroscience
Duke University
Durham, NC, USA

Will Davies
Oxford Uehiro Centre for Practical Ethics
Oxford, United Kingdom

Ken Diviney
Ryan's Rally LLC (Ryan'sRally.org)

Nita A. Farahany
School of Law
Duke University
Durham, NC, USA

Joseph J. Fins
Division of Medical Ethics and the Consortium for the Advanced Study of Brain Injury
Weill Cornell Medical College
New York, NY, USA

Jacob Gipson
Central Clinical School
Monash University
Victoria, Australia

Katherine Grichnik
Mednax National Medical Group
Sunrise, FL, USA

Valerie Gray Hardcastle
University of Cincinnati
Cincinnati, OH, USA

Jennifer Hawkins
Department of Philosophy
Duke University
Durham, NC, USA

Jakob Hohwy
Faculty of Arts
Monash University
Victoria, Australia

Guy Kahane
Oxford Uehiro Centre for Practical Ethics
Oxford, United Kingdom

Neil Levy
Oxford Uehiro Centre for Practical Ethics
Oxford, United Kingdom

Lorina Naci
Experimental Psychology
University of Western Ontario
London, Ontario, Canada

Adrian M. Owen
Department of Psychology
University of Western Ontario
London, Ontario, Canada

Barbara Pohl
Weill Cornell Medical College
New York, NY, USA

Julian Savulescu
Oxford Uehiro Centre for Practical Ethics
Oxford, United Kingdom
Macquarie University
Sydney, Australia

Caroline Schnakers
Department of Neurosurgery
University of California at Los Angeles
Los Angeles, CA, USA

Joshua Shepherd
Oxford Uehiro Centre for Practical Ethics
Oxford, United Kingdom

Walter Sinnott-Armstrong
Department of Philosophy
Kenan Institute for Ethics
Duke University
Durham, NC, USA

Rachel Zacharias
Duke University
Durham, NC, USA

1

Finding Consciousness

An Introduction

MEGHAN BRAYTON AND WALTER SINNOTT-ARMSTRONG

Ken Diviney has been caring for his adult son, Ryan, for almost 5 years. A violent assault caused Ryan's traumatic brain injury and led to a severe disorder of consciousness. Ryan's injuries, however, affect far more than just his brain; every part of the body is affected. In caring for his son, Ken is in a constant battle against injury to Ryan's fragile bones, against infections such as those of the urinary tract and bloodstream, against kidney stones and muscular stiffness. Ryan has undergone nine operations, including one that removed a third of his skull. Based on his own reading and experimentation, Ken has transformed his basement into a sterile rehabilitation center containing a therapy area, hyperbaric chamber, hospital bed with percussion vest, shower, and kitchen, where he dispenses vitamins and medications, performs mouth care every 2 hours, and administers intensive physical therapy.

Ken had to figure out much of what he wanted to do on his own, because even medical professionals misunderstand Ryan's condition. For instance, Ken was treated with suspicion and questioned about potential abuse when he brought Ryan to the doctor for treatment of a broken arm. Because Ryan is bedridden, his fragile bones are prone to breaks and fractures, but so very little is commonly known about severe disorders of consciousness and their effects on the whole body that even many care workers do not know what to expect or how to treat Ryan. Where one specialist advises a particular medication, another warns of the disastrous effects of that medication. Without much communication among the many doctors who oversee the various aspects of his son's care, Ken wades through conflicting instructions and is left with tough choices in deciding priorities for Ryan's care. The lack of shared knowledge among the many relevant people who care for his son is the highest hurdle in orchestrating proper care for Ryan.

Finding Consciousness opens with this real story of a father working to care for his son (Chapter 2) because the questions posed in this book, and their tentative answers, have real-life implications for the Diviney family. Thousands of families are dealing with these same issues. Although the exact number of patients with disorders of consciousness is unknown, it is estimated that as many as 15,000 patients in the United States have been diagnosed as being in a persistent vegetative state, and more than 100,000 others as being in a minimally conscious state (Hirsch, 2005). An estimated 5.3 million people in the United States live with some disability caused by a traumatic brain injury (Centers for Disease Control and Prevention, 1999).

Despite their prevalence, disorders of consciousness are widely misunderstood and difficult to accurately diagnose. Many people confuse these conditions:

- Coma—an acute, transitory condition in which the eyes remain closed
- Persistent or permanent vegetative state (PVS), also called unresponsive wakefulness syndrome (UWS)—a chronic or transitory condition in which the patient does not respond to commands even in the presence of eye opening
- Minimally conscious state (MCS)—a condition in which patients cannot communicate or manipulate tools but show more than reflex motor behavior
- Locked-in syndrome (LIS)—a condition characterized by complete immobility, without any required level of consciousness

Diagnosis of these conditions is discussed later, but first we need to contrast them with death. After all, a patient must be alive in order to have any disorder of consciousness.

CONCEPTS: DEATH AND CONSCIOUSNESS

To understand the new and confusing world of consciousness, it is crucial to review the history. Until the 1950s, cardiovascular, pulmonary, and brain functions were tightly woven together. The failure of one resulted in the failure of all, so the end of any single function simply meant death of the whole. With the introduction of positive-pressure mechanical ventilation, these three signs of life were pried apart. The criteria for death then needed to be re-evaluated, because a person could continue to be alive in one or two functions while a third was supported mechanically. Disorders of consciousness such as PVS were particularly confusing, because in the course of human history there had never before been a way to keep someone alive for a long period absent some signs of consciousness. Doctors discovered a brand new frontier on the border of life and death.

Jeffrey Baker from Duke University begins the discussion (Chapter 3) by laying out the broad medical and cultural history that has informed existing beliefs about death and consciousness. He looks at current perspectives through the lens of the last few centuries: the enlightenment optimism that science could restore life, the fear of being buried alive enshrined in legend by Edgar Allen Poe, and the development of more sophisticated tools and technology for physicians which widened the gulf between them and lay diagnosticians, giving physicians greater authority in defining death.

One of the most significant of these tools began with the Dinkler respirators or "iron lungs" of the polio epidemic. Endotracheal positive-pressure ventilation was soon applied to other kinds of severe conditions in addition to polio, with differing degrees of success. Some patients did not recover consciousness, but their hearts kept circulating oxygenated blood throughout their bodies. As more patients were supported by long-term respiration while remaining unresponsive, doctors began to rethink the purpose of these extraordinary life-sustaining measures.

The possibility of organ procurement further complicated the issue. If the prognoses for these long-term patients were really hopeless, other lives could be saved by transplantation of organs from these healthy bodies uninhabited by any consciousness. The specter of the hopeless prognosis prompted Pope Pius XII to condone a balancing test: If the good that the life could experience were not at least equal to the burden of the "extraordinary" interventions, withdrawal of support was acceptable, but the question of whether the patient could be considered alive or dead was one for the medical profession. Soon afterward, doctors began to declare patients who remained unresponsive over the longer term despite artificial life support to be "brain dead."

Important court cases followed. Guardian decision makers, believing that their loved ones would not have wanted to continue treatment, fought for the right to refuse unwanted medical intervention and won. Since 1990, all American citizens (or their surrogates) have had the right to refuse any treatment, even nutrition and hydration (*Cruzan v. Director, MDH*, 497 U.S. 261 (1990)). Some bioethicists fiercely debated the wisdom of this right, which was seen as passive euthanasia of those judged less valuable. They compared this situation with an imagined similar scenario for a child with a severe intellectual disability or an elderly parent in failing health: If family members believe the future of the individual to be hopeless, is denial of food (e.g., removal of a feeding tube) permissible?

When surrogates disagree with the doctors or with each other, cases become more complicated, as was demonstrated very publicly in the Terri Schiavo case. After many years of therapy and seeing no improvement, Michael Schiavo, Terri's husband, decided that it would be best to end artificial hydration and nutrition for his wife. His decision created a national frenzy with years of legal hearings, protesters holding vigils outside Terri's hospital, and children being arrested for trespassing while bringing Terri symbolic cups of water.

Cases like this raise striking ethical questions that often hinge on uncertainty. At a certain point, diagnosis and prognosis rely on probability. For prognostic purposes, PVS is considered permanent if it persists longer than 12 months after a traumatic injury or 3 months after an anoxic event. Families often find these definitions arbitrary, and the many documented misdiagnoses enable them to retain hope in the possibility of the patient's regaining consciousness. On the other hand, there remains the fear of trapping a loved one too long in an unbearable condition that is much like being buried alive.

To quell these deep-seated cultural fears, a precise understanding of death is necessary. Probably the most widely accepted definition of death was developed by James L. Bernat from Dartmouth College, along with Bernard Gert and Charles Culver. Bernat's chapter in this volume (Chapter 4) outlines the components of brain death. This biological (as opposed to spiritual) paradigm is restricted to living organisms, specifically higher vertebrate species. This paradigm maintains the ordinary, everyday meaning of death, with the only options being dead or alive: Death is an irreversible event, not a process. The death of the organism as a whole is separate from the death of its parts, meaning that the parts can die while the whole remains, and vice versa. The essence of the concept is the "irreversible cessation of the functioning of the organism as a whole. Once an organism has irreversibly lost its totality, completion, indivisibility, self-reference, and identity, it no longer functions as a whole and is dead."

If a patient is not brain dead but is alive and has a disorder of consciousness, the next question that must be addressed is, "What is consciousness?" This question is bewildering because it is impossible to capture what consciousness is if we cannot imagine what it is to be without consciousness. We always imagine through the lens of our own conscious modes. Moreover, the word "consciousness" is used in so many ways, and its many distinct meanings are so often conflated in everyday usage, that it is no wonder most people find consciousness confusing. It is necessary to carefully analyze and distinguish the various meanings of consciousness in order to diagnose and understand disorders of consciousness.

This is the goal of philosophers Tim Bayne and Jakob Hohwy in Chapter 5. One central distinction that they draw is between global *modes of consciousness* and fine-grained *states of consciousness*. They describe a mode of consciousness as a global way of being conscious; waking, dreaming, seizures, and MCS are examples of modes. States of consciousness, in contrast, are content specific; examples include hearing music or feeling pain. Bayne and Hohwy also compare modes of consciousness with "levels of consciousness" in neurology and illustrate the difficulties with classifying degrees of consciousness, arousal, or wakefulness (orientation to the environment).

These authors then propose a unique framework, seeking to describe various modes of consciousness and distinguish their permutations from each other. They allow for ordering of modes, such as perceiving wakefulness as a higher-level mode of consciousness than sleep. Still, they recognize that a complete ranking (or a complete taxonomy) may be impossible to create and would be hard to apply in practice.

DIAGNOSIS

Once we understand the basic concept of consciousness and what is lacking in these patients, the next question to ask is how disorders of consciousness can be diagnosed in individuals. How can clinicians tell whether a patient is actually in PVS, MCS, LIS, UWS, or a coma? Caroline Schnakers from the University of Liège is known for her work validating the Coma Recovery Scale-Revised. In Chapter 6 in this volume, she gives a broad overview of the traditional bedside scales—the Glasgow Coma Scale, the Coma Recovery Scales, and the Nociception Coma Scale—and their various uses and challenges when detecting signs of consciousness. Because misdiagnosis has such serious consequences for the type of care available to the patient, including pain management and end-of-life decisions, Schnakers emphasizes reliance on objective, standardized, sensitive criteria and adherence to strict administration and scoring guidelines in order to avoid misdiagnosis.

Errors do happen, however. Error can be introduced when the examiner samples too narrow a set of behaviors or does not properly define the criteria for intentional responses. The examinations could be too infrequent or the time windows too short, missing the desired behavior. Getting an appropriate sample is important. Timing matters because the patient's ability can be affected by many factors that would have an effect on anyone's ability, such as pain or discomfort, impairment from medications, or environmental disturbances. Poor conditions may decrease the probability of observing the signs of consciousness. To reduce such errors, Schnakers emphasizes reliance on the Nociception Coma Scale and careful monitoring to ensure that pain is being properly managed. As her overview closes, Schnakers notes that advances in neuroimaging have been able to offer other insights into patient consciousness and should be considered as a complementary tool to distinguish patients with PVS from those with MCS.

These neuroimaging techniques add much more than merely a tool for distinguishing types of patients, however. Adrian Owen at the University of Western Ontario has really changed the game in understanding what is happening in some patients with disorders of consciousness. These new technological capacities, which were once only in the realm of science fiction, are summarized in the chapter by Owen along with Lorina Naci, also at the University of Western Ontario (Chapter 7). They take on the very real possibility that neuroimaging could be used to determine thoughts, intentions, and other mental states directly from brain responses without needing to rely on overt physical action.

Patients who had been classified as being in PVS for years have been studied with the use of functional magnetic resonance imaging (fMRI) to demonstrate the presence of conscious awareness and complex mental function. Revealing sessions have shown some patients' repeated brain responses to motor imagery tasks (e.g., imagining swinging a tennis racket) and spatial imagery tasks (e.g., imagining in detail walking through the rooms of your home); the responses clearly engage brain regions that are very close to the brain responses of healthy control subjects performing the same tasks. This difference allows patients to answer yes-or-no questions by mentally engaging in the various tasks in order to indicate different answers. Other mental tasks have also been effective in repeatedly generating consistent responses. Examples include answering questions about biographical information unknown to the researchers, such as the patient's father's name or a place where the patient had vacationed—answers that were subsequently verified.

These tools are supposed to do more than just demonstrate the presence of awareness in patients with disorders of consciousness. fMRI (and EEG in a separate report) can allow patients to engage in an activity (e.g., answering questions) that requires language (i.e., understanding the questions and the instructions for answering) as well as self-knowledge (e.g., their own father's name) and memory of the past (i.e., because they had not seen their father or been told his name in the years since the brain damage). These capabilities are amazingly advanced for anyone diagnosed as being in PVS. Some observers have concluded that such patients have some potential for competence in decision making, although this idea is controversial.

A healthy dose of skepticism in provided in the chapter by Will Davies and Neil Levy (Chapter 8). They argue that the experiments by Owen and others do not succeed in proving any significant intention, agency, or consciousness.

The ability to follow commands and answer questions shows only that these brain-damaged patients respond to external stimuli. These methods do not (yet) reveal any endogenous intentions of internal origin. They also do not (yet) reveal the kinds of connections among intentions that are required for plans, interests, or preferences about how life will proceed. Davies and Levy claim that patients who pass these recent tests still might have no more consciousness than patients with a different diagnosis: extreme akinetic mutism. They conclude that these tests do not demonstrate what is relevant to the moral status of being a person. That conclusion leads directly into the next set of chapters.

ETHICS: WHAT MATTERS?

The natural question to ask next is, "Given what we know and do not know about these patients, what should we do?" In particular, should we keep these patients alive or let them die? Any answer is bound to be controversial, but we can at least understand the issues better by considering conflicting arguments.

Jacob Gipson, Guy Kahane, and Julian Savulescu at the University of Oxford begin by outlining a general framework for addressing ethical issues (Chapter 9), applying Beauchamp and Childress's principles of autonomy, beneficence, nonmaleficence, and justice. Then they report a survey of how everyday people rank these principles and reach overall judgments about whether patients with various disorders of consciousness should be allowed to die. The survey responses vary in fascinating ways depending on whether the question is asked abstractly or concerns an actual case and whether the question is about other people or asks whether the respondents would want to die if they were in such a conditions themselves. Gipson, Kahane, and Savulescu close by arguing that popular opinions about these matters have indirect relevance to normative issues regarding what should be done in specific instances.

Instead of appealing to surveys, most philosophers appeal to arguments and theories. The rest of this section of the book includes three such philosophers who suggest in various ways that both practitioners and the general public need to revise much of what they think about the ethics of these cases.

Joshua Shepherd at the University of Oxford begins his chapter (Chapter 10) with the assertion that consciousness is morally significant in itself, bringing with it its own ethical complications. Shepherd's focus is on MCS, because patients with MCS show some degree of consciousness, even if their episodes of consciousness are unstable and intermittent. In order to specify more precisely what is morally important in cases of MCS, Shepherd discusses two distinct notions of consciousness: *access* consciousness and *phenomenal* consciousness. The possession of access consciousness entails abilities to use information for reasoning or controlling behavior. In contrast, the possession of phenomenal consciousness entails that there is something "it is like" for the possessor (such as what it is like for a patient in a MCS). Shepherd argues that, if there is something that it is like to be in a MCS as opposed to a PVS, then MCS has a moral significance that PVS lacks, because phenomenal consciousness is required for the possession of some degree of subjective well-being.

However, if the MCS patient, when competent, clearly expressed a desire to die, then a difficult moral conflict may be posed between the values of autonomy and well-being. If autonomy is given priority, so that the desire to die is granted, then well-being and present and future enjoyment might be harmed. But if well-being is given priority, so that the remaining quality of life is preserved, then the patient's right to determine his or her own life is violated. Shepherd finds that both autonomy and well-being need to be considered, along with other values such as distributive justice. That conflict of values explains why such cases are so difficult.

In contrast, Jennie Hawkins at Duke University (Chapter 11) asks whether the crucial question is what the patient wanted, or believed he would want, when writing an advance directive. At that past time, the patient had not yet experienced a disorder of consciousness. Therefore, instead of asking what the patient wanted, perhaps we should focus on what is in the best interests of the patient.

Hawkins distills the debate down to two questions that she believes are at the heart of making the ethical decision: whether patients suffer and whether they derive benefit from their lives. She ends up defending the choice to allow patients in PVS or MCS to die.

Hawkins relies on the concept of prudential value, rather than welfare or well-being, because there are times when people are struggling and suffering but continuing life might still be the best option. We need to look at which choice would be good relative to all other available options. For life to have value, in Hawkins' view, one must have the capacity to form relationships and to value. Intermittent awareness and lack of communication would lead to such loneliness as to make death preferable.

For those with a disorder of consciousness, Hawkins believes that the benefits derived from life are small enough that suffering would clearly tip the balance toward death. Even if there were no suffering, she says, we ought not to assume that life is automatically preferable. Not having reason to die is not the same as having a reason to live. If a baby were born with neither consciousness nor capacity to develop it, then the child will derive no benefit from life and death is no harm, so the burdens of raising the child can tip the balance toward death.

Opponents are bound to object that patients should choose for themselves, because recent technological developments enable patients to communicate through fMRI or EEG. However, Hawkins questions the value of these messages. There is no way, she argues, to assess the quality of the decision making in such cases or the process behind it. Mere expression of preference does not sufficiently establish that the patient is competent to make decisions. Gaps in the scientific findings lead to doubts about how much to credit the communication that is currently possible.

Whereas both Shepherd and Hawkins assume that consciousness matters to morality, Valerie Gray Hardcastle at the University of Cincinnati (Chapter 12) questions this common assumption. Instead of consciousness, Hardcastle focuses on pain and uses studies of subliminal processing to argue that consciousness is not required in order to perceive pain. She reinforces this claim with evidence that there is a widely distributed neuromatrix for pain reception in the brain; it is not a highly localized process.

Hardcastle also questions whether consciousness determines how much we should care about a living being. Apparent expressions of pain from PVS patients, such as "grimacing, posturing, crying, even racing heartbeats and hormonal fluctuations," could merely be autonomic and unconscious responses. But does that mean they do not matter? Moreover, unlike PVS patients, MCS patients have neural responses to pain that are similar to those of healthy controls, including responses in the thalamus, primary and somatosensory cortex, insula, and cingulate, as well as "the co-activation of specialized sensory cortices and frontoparietal areas." This suggests conscious perception of the nociceptive stimuli.

Hardcastle posits that we do not know what it is like to be in such states of limited consciousness. In particular, it is difficult to establish the negative conclusion that any such patient really has no consciousness or perception of pain. And, as Schnakers and Owens demonstrated, there is a high number of misdiagnoses. For these reasons, an assumption that patients who have been diagnosed as being in PVS or MCS lack pain or consciousness seems reckless because it could result in extreme and extended suffering. The potential for awareness should prompt medical personnel to err on the side of pain management. We should assume that pain is possible and act to treat it.

PRACTICAL ISSUES: LAW AND MEDICINE

Each of the different perspectives articulated in this volume could arm policymakers with a good deal to think about. Some contributors who come at these issues with an eye toward making policy close this volume with practical considerations.

Nita Farahany and Rachel Zacharias from Duke University address many of the same questions about consciousness, death, and pain from a legal perspective (Chapter 13). They introduce a theoretical framework by which to define legal life and death in terms of consciousness. The legal standard for death has implications for tort, estate, organ donation, and criminal law (e.g., whether a crime is homicide or assault). They call this framework "the legal circle of life."

Farahany and Zacharias suggest that a legal standard should hold based on its purpose, regardless of the technologies available now or in a few years. Consciousness has had and continues to have value in the life of the individual. The belief that life ends with the end of consciousness is controversial, but not so controversial as the application of this principle to the abortion question. When viability also depends on technology, consciousness may be the more important question. Many lawmakers have tried to use nociception as the guideline for when a fetus begins to feel, or be aware, or be conscious, but responses to unpleasant stimuli are not the same as processing pain. Consciousness of pain "requires a somatosensory cortex," and "that signal must come through a functioning thalamus." Even though there may be cortical activation, there cannot be conscious perception of pain. These issues lie at the intersection of law with neurology and philosophy.

In the final chapter, Joseph Fins brings us back to the issues raised by Ken Diviney at the beginning of this book. Many of the real decisions about treatment fall into the hands of the patient's guardian. Guardians aid patients in many essential ways. They advocate, protect the patients' interests, seek treatment in facilities that might not have offered it, and affirm the right to care. Still, there is a potential for infringing on a patient's civil liberties or limiting self-determination, because guardianship is assumed to be permanent. Because increasing numbers of these patients will regain consciousness and independence, however, it is essential to structure guardianship to protect patients both when they are incompetent and when there is a possibility for regaining competence.

For example, how does guardianship come to an end when it is no longer needed? Most guardianship situations either have a specific termination date (e.g., when the ward reaches 18 years old) or are considered indefinite. If the patient can contribute to decisions about his or her own care, an overzealous guardian may hamper the patient's agency. When a guardian is appointed by a court from outside the family, his or her decisions might not reflect the patient's true wishes and could strain relations within the family. A distributive justice question also comes into play when long-term care and rehabilitation facilities require guardians for all their patients. Patients with no relatives and fewer resources may not be able to secure a guardian and may have worse outcomes than more connected and wealthy patients.

Fins recommends that close family members be privileged in guardianship because they tend to have the greatest knowledge of, and respect for, the wishes of that individual patient. Guardianship is especially important when there is a need to establish a trust and prepare for spending over the life of the patient. Diagnostic errors are always possible, and the outcome may differ greatly from the expected course of the patient's disorder of consciousness. For this reason, guardians must, above all, be wary of the existing cultural nihilism toward those with disorders of consciousness. Any guardian who neglects to aid recovery of consciousness and ability is abusing the position and infringing on the rights of the patient. Much like Shepherd, Fins speaks of weighing the expressed interests of patients alongside their present and future well-being in order to properly represent those patients who do not have the ability to advocate for themselves.

For the tens of thousands of guardians like Ken Diviney, who are trying to work through the ins and outs of how to help loved ones with disorders of consciousness, and for those of us who are looking to understand what our own consciousness means, these real questions hold important meaning. They also raise profound issues for courts and policymakers and will stimulate much debate both inside and outside of academia for decades to come. We all might end up in with a disorder of consciousness or with a close relative or friend who has a disorder of consciousness, so we all need to think through these difficult questions.

REFERENCES

Centers for Disease Control and Prevention, National Center for Injury Prevention and Control. (1999.) *Traumatic brain injury in the United States: A report to Congress.* Atlanta: CDC.

Hirsch, J. (2005). Raising consciousness. *The Journal of Clinical Investigation, 115*(5), 1102.

2

Discussion with a Caring Father

KEN DIVINEY AND KATHERINE GRICHNIK

The following is an edited, transcribed audiotape of an interview performed during the Finding Consciousness workshop at Duke University, January 24, 2013. The participants are Katherine Grichnik, MD, MS, FASE (KG) and Ken Diviney, father and caregiver from Ashburn, Virginia (KD).

KG: This is Ken Diviney. . . . Ken's son, Ryan, is . . . 23. Three years ago, he was walking to a convenience store behind his college house and there was some verbal exchange that took place between Ryan and his group of friends and a second group of friends—of young people over sports. . . . Actually, why don't you tell the story? Because I think you'd be better to explain it. . . .

KD: So the short end of it is that he was blindsided by a sucker punch, went to the ground, hit his head on a raised grate, was unconscious, and another kid kicked him. So his injury is diffused.

Before I get into all that, what you guys are doing here today . . . is right on the mark. Now, we need to bring you all together and get a solution because I see there might not be the connection between the philosophical and the legal and the neurological and the countless other disciplines that are involved in brain injury. Somewhere, that's got to come together so it can be practically applied at the patient level. Now I've been able to track with all of you, from the wonderful keynote last night and the Coma Recovery Scale this morning. I have sleep studies I can pass off to you.

And that's one of the problems of being the caregiver in all this. I don't have a centralized source. So I'm going out, I'm trying to get information because, if there is a medical breakthrough . . . or some sort of discovery, I need this data because you guys need this data. And that's the only reason. And it doesn't matter what I see, because it's not real until you see it. . . .

KG: . . . I have a set of questions that I had shared earlier and I was going to walk through. But I think it would be more important that, as I ask the questions, for you to participate and ask as well. Because this conversation can go anywhere we want. I don't want to prescribe it, and I would like us to have a two-way conversation.

So, first I want to ask about making that transition from an interoperative setting . . . to just taking care of Ryan at home. When he was first injured, what did you understand from the medical team about what was likely in the future? How were they able to talk to you? Or what could you hear?

KD: Exactly. Because when you walk into that sort of situation, I don't know your language. I think I've learned it well now. So what it got down to is one doctor saying he's in grave condition and another one saying [otherwise]. . . .

So finally I got a hold of Julian Bales, who was at Ruby Memorial. He's with the NFL now. . . . We actually are friends with Sanjay Gupta, so he called Bales and brought him in. And my only question to Bales, was: Is there no reason to hope? I didn't care about anything else, and that answer was going to dictate where we went next. Because the summer before, my son, my wife, and I were all talking and it all came out of my wife saying, Did you ever fall asleep and you wake up and you feel like you're in a coma? And we got on that sort of discussion. My son specifically said you have to do everything if I'm ever there. So I knew, going in. But if there's no hope, then there's no reason.

KG: So Dr. Fontes just said we always try to convey hope. Is that reasonable?

KD: It's reasonable, as long as it's realistic.

AUDIENCE MEMBER: I don't mean to interrupt.

KG: No, please. We would like a discussion.

AUDIENCE MEMBER: I have a hard time dealing with the whole process when I'm the intensivist—and I have clinicians, physicians, caretakers—and I'm offering hope when I feel strongly there's no hope. What does one do at that point? Clearly you don't want to complicate the situation more. But it is not to say that they're wrong either, because they've told you there was one patient that he or she had so-and-so who was comatose for 2 months in the ICU and they walked out of there. I've had such experience.

KD: Remember the mine collapse in West Virginia that was 4 years ago? Julian Bales' patient was the one that walked out. He was the one that walked out of there. What do you do? When you don't have the information, you default to the decision that we're going to continue. The default decision is never "Let's go ahead and terminate now." It's never that.

KG: . . . So the decision was made to go ahead and operate and to relieve the pressure and allow his brain to swell in reaction to the injury. But that was clearly an operative procedure that you had to have consented for. And so you gave consent, I think maybe knowing that that triggered the rest of this series of events.

KD: It was the decision of a lifetime.

KG: But you probably didn't quite know where that was going at that point.

KD: Oh, no. There's no way. I'm telling you this has to be the worst injury that a person can have inflicted upon them because it hurts them everywhere. It touches every part of the body, right down to the cellular level. I mean UTIs and it's a constant battle. Things that you wouldn't even associate with it. . . .

KG: Again, I just want to ground us in what reality is now. So we were talking about what happened at the time that Ryan was first injured. He's still in the hospital transitioning to the skilled-care facility. Just give me a snapshot. Has he woken up today?

KD: Well, I think I'm kind of in agreement with everyone here. I don't know. It's possible. There's been imaging that shows brain activity and we use—every 3 months we do QEEGs, MRIs. Things light up. Ask him questions, this and that. Is he awake? Not like you and I are. Sometimes he's purposeful. Then it goes days, weeks, months, and I'm like, oh, maybe I was just imagining that.

AUDIENCE MEMBER: Is he getting treatment?

KD: He gets a lot of treatment.

KG: You do it.

KD: Yes.

AUDIENCE MEMBER: Is he getting any medications?

KD: He does. Well, we use a lot of off-label type of stuff. Depending what sort of region of the globe you come from, it's tiracetam/piracetam. We have amantadine, bromocriptine, cardidopa, levodopa, Ritalin . . .

AUDIENCE MEMBER: So you're managing all the dosing?

KD: Yes, exactly. And then things to boost the immune system, nutraceuticals. At one point, I counted. It's like 40.

KG: Every day. Well, not all at one time. Spread out.

KD: Yes, spread out, spread out throughout the day. And then I retract him from that and see what happens. You know, give him a holiday from this one. It's very complicated.

KG: So, I'm going to bring us from today back to the skilled-nursing facility. When you transitioned to that time period, was there any one person or setting that was particularly helpful—or not—in helping you to understand, to project out what your life was going to look like here?

KD: No.

KG: What was your understanding at that point of what was going on?

KD: Well, that he was medically stable.

KG: Well, we've used that term, haven't we?

KD: I'm telling you, medically stable was not what I envisioned at that point because when he went over to this rehabilitation facility, he was still having neurological storms and, second night in there, his temperature going up and up and up. We can't get a doctor in. All we want to do is ask: Is this normal? Is this something that happens? How should we deal with it? I mean, I've never left him and we could not get a doctor in the room. You know how we resolved that problem? We called 911 on the ICU.

Yes, that was an interesting night. And we had plenty of doctors in the room after that. They were all interested.

AUDIENCE: [Laughter].

KD: So then the center we took him to was Shepherd's Center in Atlanta, which is very good for spinal cord injury. And they're just starting to get it together for the brain injury. But they were the only facility on the East Coast that had an ICU in their rehabilitation facility. And in their facility was a tunnel to Piedmont Hospital. Without that Center, we would have had nowhere to take him. We didn't know what to do. So we considered that a short-term relief. He stayed there until February, and every 10 days they would come in and say, well, we're going through the insurance process, duh-duh-duh-duh. . . . And they got to February and I said stop it. Just stop it. We're going home. I know what I'm doing now. I can give a clinic on respiratory care. And if you need, I can suction you.

AUDIENCE: [Laughter].

KD: So we took him home and we read a study about a new treatment. By the way, I've probably read everybody's studies in here. And I probably emailed you too. So we took him to Kessler Center in New Jersey because they were doing some protocols that we liked, and they came out in an article in *Newsweek* that was called "Waking the Dead," I believe it was. They were using some interesting things. But by September nothing was really different, so I brought him home.

I've been tending to him ever since. Converted the entire basement. And fortunately we had a fairly big house, so I had about 1600 or 1800 square feet I could work with. Put in a huge bathroom with a shower. I tell people you could have a Roman orgy in the shower. It's just like that. Basically a hospital room, a therapy area, a prep area with a kitchenette, and then just an area to get away from everything. Had to put all new ramping in. We probably bought so many yards of concrete that we could have done two tennis courts with the amount we bought. We brought in—I know it's highly controversial—hyperbaric oxygen. Bought the chamber. I have a chamber in my house. He gets it on/off—30 days on, 30 days off—3 hours a day sessions. Functional electrical stimulation on a bicycle. Every morning, I hook him up to electrodes. His muscles are doing the work. It's involuntary, but they're doing the work.

KG: Who helps you? Who is the physician who helps you? One of you asked how many medications and who manages that. I mean, you seem so capable. Now, you've got to have a partner in this.

KD: No. The way it works is neurologists will prescribe the ones that they prescribe and then the urologists, you know, we want to put him on low-dose Macrobid, prophylactically we want to give him oxybutynin to take the muscle tension off his bladder. But then this one could affect brain awareness here. So, what I do is I manage all the doctors. You've got urologists, you've got neurologists, you've got all the "-ologists" out there. And then you've just got to make decisions on your own. They gave him a baclofen pump, which helps with spasticity on some patients. And he was getting something like 700 micrograms to start this about a year or two years ago. I thought, you know, I'm just going to start dropping this. And nobody questioned me. So, slowly we started weaning him down. The pump was removed in October.

KG: That's less infection.

KD: Exactly.

KG: Jeff, you had a question?

JEFF BAKER: Some of my patients who have been in very tough situations, their families will express how, when they're in the hospital, they can't really tell who is in charge. . . . Sometimes one person doesn't take charge when you're in the ICU.

KD: One time his bone density was going down, and fast. And he actually did completely break the ulna and fractured the radius during therapy. And I take him to the hospital. And this is the crap that we have to deal with as caregivers. I take him to the hospital after calling up my physician and they're saying, yes, you should probably take him in. And pretty soon some guy is walking in asking me a bunch of questions. And then it dawns on me. I'm like, wait. Where are you from? Who are you? And he was adult protective services.

KG: Oh, my gosh.

KD: So, this is all the fallout from that. His bone density has decreased, he's in physical therapy, and he sustains an injury. And all of a sudden the caregiver, the best caregiver that any patient could have, is suspect. That's being pissed off. But you can't act that way because then you feed into it.

KG: Oh my goodness. So, you've actually touched on some things that I wanted to bring up that I think are sensitive. And it's not only about caring for your loved one in this situation where it's 24/7, but it's also the financial pressures that go along with this, and I'm going to say the emotional pressures.

KD: Let me talk about financial pressure. Brain injury touches every part of the body. Does anyone disagree with that? Insurance doesn't understand that. So, I go out and I try to get a therapy table and they think, well, you can just do that in bed. But I have him up at 5:00 a.m. every morning and he doesn't go to bed until 9:00 p.m. They don't understand that if you leave him in bed, it just contributes to the injury. And this is what happened. I was looking at some of these pictures of the contractures. My son is as loose as can be. He wasn't at one point, when I first brought him home. If I were strong enough, I could have lifted him up like this [demonstrating lifting his son up like a board], and he would have been straight across. So, I wasn't willing to accept that he was going to be that way because it was painful for him. And it doesn't matter if he was consciously or unconsciously hurting. It hurt.

KG: It hurts you too, though.

KD: Oh, it hurts me too. So, I just said the hell with it and I just started stretching him, stretching him. And it wasn't like the type you get in the hospital where they come in and dink around there a little bit and leave. I stretched every joint. Had him hooked up to a pulse ox. I would see his pulse—and this time he was kind of tachy—but he would go up to 150 beats per minute, and I would have to wait and let him come back down and I would work the next joint. And the next joint. It took a year, and maybe over a week you would gain a millimeter of range a day. But eventually I got him to the point where he's loose and he's comfortable. And his pulse reflects that.

One of the topics that came up today in your discussion, was reliance on the equipment to tell you the patient's status. Allow me to share a personal experience. It was the first time I ever walked out of his room, my wife calls me and says you've got to get back here. They had hooked him up to an external defibrillator just in case. So I come running back into his room and I look at him and it's shocking him. Boom! You know, every so often. Boom! And I walk over to the doctor, I said, has anyone actually taken his pulse? Well, no. So, it was normal. But in the meantime, he's getting defibrillated.

KG: So, in my ICU doctor hat, I always think that in this situation, one of our worst problems is infections. Because people don't die from the brain injuries, they die from the infections. Often. How do you deal with that?

KD: Well, I closely monitor who comes into my house. It's close to a sterile environment. I mean, I've even put those UV lights in my filter, in my furnaces, to clear it out. But there's also some precautionary things that I've read through research. You know, pneumonia is always a problem with a person in this condition. And to combat that, we got him a bed that has a percussion therapy. We have a cough assist in our house. And I brush teeth or swab his mouth every 2 hours. He hasn't had pneumonia since December 2010.

KG: Which is remarkable in 2½ years. Yes.

AUDIENCE MEMBER: One thing that I think would be extremely difficult about a parent taking care of a child is constantly wondering what's going to happen when I'm gone. I wonder to what extent you've battled with that. Does he have brothers or sisters who can take care of him when you're gone? Do you worry about that?

KD: Oh, all the time. All the time. He has a sister, but she has a life. And I tell her, you're not your brother's keeper. You're not. It's not your responsibility.

The responsibility, or the accountability, lies with the people who did this, and we're still going through that whole process. So we've set up trusts, and we've taken care of all the things that we can take care of. And it's not even so much if I die. It's . . . what if I become . . .? You know, I'm 48. I can't even believe I'm still going at 48. But, what happens when I'm 70? You know, his longevity is probably going to be more than other people because his care has been extraordinary. He's in pristine condition. If you walked in, you wouldn't even know. You'd be like, hey, get up. And that's, you know what, that's hard to look at too, when you look at him and you're like, what's going on?

KG: Most of us think that when our loved ones get into the situation, that the doctor knows best, or the health care team can take care of this, or I'll go to a skilled-care facility and they'll get PT and OT and respiratory therapy and all of that. But, how did you learn to be such a strong advocate? You took over what most of us would think would be someone else's responsibility. I'm not being trite. It's just, you're remarkably strong.

KD: Thank you for that. But, you know, my wife and I, we made a decision, which is now 24 years ago when we conceived him that no matter how he was born, no matter how he progressed, we were going to continue on with this. So let's say he would have been born with some sort of long-term disability—we accepted that obligation at that time. Two decades removed doesn't change that. We have an obligation to our family. And, you know, when I would walk down the hall of these rehab centers he was in—and I never left my son—I could see patients in better condition that nobody was visiting. And you know what? They didn't really have a chance.

KG: That's true. So, another sort of, perhaps, tough question. Is there anything that you think a physician, a healthcare provider, anybody could have done to make a difference in the outcome? So, in a roundabout way, I'm asking: Do you think anybody made a mistake?

KD: Oh, sure, people make mistakes. But, they were done with the right intention. And it's all hindsight now. If I could have gone back to that night, I would have now had the knowledge to ask, well, why aren't we chilling him down? Why aren't we giving him estrogen? I now know all these things that I could have asked. And, who knows? At that time, it was just kind of like, oh, yeah, this might be something that helps. Yes, everybody makes mistakes, and I've made mistakes as well. But you move on. It's what you have to deal with. And sometimes your best intentions aren't the right course, so you have to correct and make it right.

PART I

Consciousness

3

The Geography of Unconsciousness

From Apparent Death to the Minimally Conscious State

JEFFREY P. BAKER

The essays in this volume explore the implications of recent neuroimaging studies suggesting that some patients—perhaps *many* patients—who currently bear the diagnostic label of "vegetative state" may in fact possess at least some degree of awareness and responsiveness. They examine questions concerning how we should think about, and care for, persons in a mode of existence that has only recently been given a name—the minimally conscious state (MCS).

Although recognition of the MCS raises scientific questions that in some ways are quite new, it also represents the latest act in a much longer story: the mapping of what might be called the geography of profoundly impaired consciousness. This history is usually traced back to the rise of intensive care technology in the 1950s and the emergence of a growing number of patients who remained dependent on mechanical ventilators and feeding tubes. Although the 1950s did mark the beginning of the modern bioethical debate regarding states of consciousness, it is worth remembering that the public has been fascinated with the borderlands of consciousness and the disputed territory between life and death since at least the 18th century.

COMA AND APPARENT DEATH

Our word "coma" is derived from Greek *koma*, a term used in the Hippocratic corpus to indicate a state of deep sleep. As physicians in the early modern period became more interested in nosology (the task of naming and classifying diseases), they developed more precise language to delineate specific states of altered consciousness. The English physician and pioneer of neurology, Thomas Willis, distinguished coma (heavy sleep) from lethargy (pathological sleep), carus (deprivation of the senses), and apoplexy (a still more profound state of unresponsiveness). Much discussion in the 18th century sought to extend this kind of classification system and identify the causes of these various states of unresponsiveness (Koehler & Wijdicks, 2008).

Of relevance to our present discussion are those states of unawareness with names such as "suspended animation" and "apparent death" that appeared to straddle the line between life and death itself. Physicians were fascinated by rare but well-attested stories such as that of the servant Anne Greene, who literally woke up on the dissection table after her hanging in 1650 (Watkins, 1651). Starting in the 1740s, a genre of treatises began to appear proposing means by which the "apparently dead" might be restored to life. They describe an expanding array of resuscitation techniques that included the employment of mechanical bellows to inflate the lungs, applications of electric shock, and other approaches (more obviously rooted in 18th century physiology) such as bloodletting, tobacco enemas, and stimulatory vapors. The similarity of these techniques to modern cardiopulmonary resuscitation (CPR) should not be overstated; physicians had no way to truly intubate a patient, and the electricity administered via a Leyden jar was not a prototype of a modern defibrillator. Yet without question, there were occasions when patients appeared to have been brought back to life by such interventions. It is particularly easy to imagine that hypothermic drowning victims, so cold that their pulses or respirations could not easily be detected, could have been revived through warmth and stimulation (Eisenberg, 1997, pp. 55–73).

The story of how resuscitation techniques and ideas about "apparent death" evolved in the 18th and 19th centuries cannot be told in detail here, but it is worth highlighting three themes that anticipate the modern debate over disorders of consciousness. The first was the emergence of an optimistic Enlightenment faith in the power of science to reclaim life from apparent death. Whatever doubts we may have about the efficacy of 18th century resuscitation techniques, they generated widespread excitement at the time. No less an authority than John Hunter, perhaps the most renowned British surgeon of the entire century, secretly attempted to revive a hanging victim after his execution (although without success) (Moore, 2005, pp. 185–198). Philanthropic "humane societies" sprang up throughout Europe and North America to popularize and promote resuscitation. London's Royal Humane Society, founded in 1774, provided free lectures, set up fully-equipped rescue stations, trained layperson assistants, and handed out thousands of pocket-sized emergency instruction cards. Processions of victims saved through the Society's efforts marched through London, celebrating humankind's attainment of a power previously reserved for God (Lathrop, 1787, pp. 18–23; Struve, 1803, pp. 5–13). Such phenomena anticipated today's "medical miracle" stories recounted so often in popular media, as well as the predilection of television medical dramas to depict CPR as successful far more often than takes place in reality (Diem, Lantos, & Tulsky, 1996).

A second theme pointed in a very different direction. If the recognition of death was no longer obvious, was it not possible to be declared dead by mistake? Samuel Hawes, who had founded the London Humane Society, also wrote a book on premature burial, using the Society's case records to argue that only putrefaction offered a reliable indicator of death (Hawes, 1780). According to historian Martin Pernick, popular fear of premature burial became much more widespread in 19th century Western culture than is generally appreciated, and it did not recede until near the end of the century. Popular writers such as Edgar Allen Poe tapped into it. Government laws and churches began to impose waiting periods between the declaration of death and disposal of the body, and some cities went so far as to provide mortuaries allowing observation of the body in an open casket. One enthusiast developed a signal device to be placed in the coffin on top of the body, whereby any chest movement would turn on a light, open an air duct, and ring a bell for half an hour. It was widely promoted by the British Society for the Prevention of Premature Burial, which was formed in 1894 by anti-vaccination leader, William Tebb (Pernick, 1988, pp. 17–74).

The third and final theme is that of how physicians gradually established their own authority as the arbiters of defining death. Diagnostic technology played an increasingly important role in this process. It began with the rise of physical examination techniques such as auscultation in postrevolutionary Paris hospitals; physicians argued that the stethoscope could detect a heartbeat too feeble to be noted by the undiscerning examiner. Many more tools for determining death and assessing unresponsiveness emerged later in the century, ranging from medical thermometry and the assessment of pupillary reflexes to the description of the postural reflexes associated with severe brain injury (Pernick, 1988, pp. 37–47; Powner, Ackerman, & Grenvik, 1996). Although these proved to be effective for diagnosis of cardiorespiratory death and brainstem injury, assessment of the higher brain—the cerebrum—was more elusive. Many believed that the electroencephalogram (EEG), first applied at the bedside in the 1920s, would provide a window into the workings of the mind. Instead, the complex patterns of brainwaves seen on an EEG report, tantalizing as they were, found their main clinical application in the diagnosis of epilepsy. Research studies using the EEG nonetheless did shed indirect light on how consciousness worked, highlighting, for example, the role played by the reticular activating system in the pontine and midbrain space in sustaining alertness and mediating the content of consciousness (Moruzzi & Magoun, 1949).

BRAIN DEATH

Despite all of the interest in characterizing and attempting to rescue patients in unresponsive states, a patient in deep coma before the 1950s could not remain in that state for long. Recovery was only possible if normal breathing was preserved and, just as importantly, the patient's airway reflexes remained sufficiently intact to protect against pneumonia. This situation changed with the rise of positive-pressure ventilation. Negative-pressure Drinker ventilators, remembered in popular culture as "iron lungs," which had been used for polio victims since the 1930s, were effective primarily for conscious patients with neuromuscular paralysis (Markel, 1994).

Positive-pressure ventilation, which allows more direct control of airway secretions through an endotracheal tube, was confined to the operating room until a devastating polio epidemic struck Copenhagen in 1954. The crisis overwhelmed the available supply of Drinker respirators, prompting anesthesiologists to come out of their operating rooms, perform tracheostomies, and commission medical students to keep the patients alive with a bag and mask. The experiment was successful, prompting the invention of positive-pressure ventilators as a better solution than sleep-deprived medical students. After polio rapidly declined following the introduction of the Salk vaccine in 1955, the new life-support technology was applied to patients with ingestions, pulmonary disease, and coma (Reiser, 1992; Snider, 1989). Many patients recovered. Many others, however, never regained their previous level of functioning. And a few remained permanently dependent on the ventilator.

Physicians began to face increasing numbers of unconscious patients sustained by artificial ventilation, first on the wards and then in the specialized domains that would become intensive care units. At Massachusetts General Hospital, for example, the number of patients annually requiring ventilator support for longer than 24 hours increased from 66 to 398 between 1958 and 1964. Attending physicians from this time later recalled facing many dilemmas in assessing these patients' consciousness and having conversations with families that sometimes included recommendations to withdraw support (Belkin, 2003, pp. 343–346). A variety of new terms emerged to describe the growing class of unconscious patients inhabiting the wards and intensive care units of 1960s hospitals, including "coma vigil," "apallic syndrome," and "neocortical death" (Bauer, 1987). French investigators coined the term *coma depassé* ("beyond coma") to describe an apparently irreversible state of unconsciousness characterized by a flat EEG tracing which correlated with profound degeneration of brain tissue on autopsy (Mollaret, Bertrand, & Mollaret, 1959).

Recognizing they had entered uncharted moral territory, a delegation from the World Congress of Anesthesiologists decided, in 1957, to take advantage of the meeting's location in Rome and interrogate no less an authority than Pope Pius XII himself. Was there ever a point, the society's president asked, beyond which an unconscious patient whose circulation could be maintained only by artificial respiration should be considered "dead"? The Pope replied that the answer did not lie within the competence of the Church and tossed the question back to the medical profession. He also affirmed that even if such patients were considered to be alive but with a hopeless prognosis, it was morally consistent with Catholic teaching to withdraw mechanical ventilation or any other "extraordinary" form of medical support if the burden of the intervention outweighed any possible good (Jonsen, 2008).

The ethical problem of end-of-life decision making for patients on mechanical ventilation soon intersected with a second question: When could a patient be considered "dead" for the purpose of organ procurement? By the mid-1960s, advances in kidney transportation and immunosuppression led surgeons to see cadaver rather than live organ donors as the best solution for the growing numbers of patients who were dependent on dialysis. Some European surgeons began to remove kidneys from patients in *coma depassé*. They argued that irreversible damage to the central nervous system was equivalent to physiological death, permitting the removal of organs from a body that could essentially be regarded as a cadaver. Surgeons attending a 1966 CIBA Foundation Conference on Transplantation showed great interest in developing criteria for a new definition of death, although some recognized the potential for controversy. Even Dr. Thomas Starzl, on his way to becoming one of the most renowned pioneers of transplant surgery in the world, questioned whether his team could "accept a person as being dead as long as there was a heartbeat" (Wolstenholm & O'Connor, 1966, p. 157).

Both of these clinical problems—end-of-life decision making and organ donation—set the stage for the well-known 1968 Harvard report defining brain death. It was created by an ad hoc committee of Harvard faculty, chaired by anesthesiologist Henry Beecher, that included ten physicians, a lawyer, a theologian, and a historian of science. The report framed its mission in terms of defining "irreversible coma as a new criterion for death." A patient in deep coma could be considered "dead" if three (or ideally four) of the following criteria were satisfied, as assessed by two physician examinations at least 24 hours apart: complete unresponsiveness, lack of spontaneous breathing, absent brainstem reflexes, and a flat or isoelectric EEG. The

EEG criterion was not mandatory but was considered "of great confirmatory value," reflecting continuing disagreement over its meaning as well as the fact that EEGs were still far from universally available in American hospitals in 1968. These clinical criteria equated brain death with the absence of basic brainstem reflexes, not just absence of consciousness or higher cortical functioning. The report contained only one reference, the statement of Pope Pius XII from a decade earlier, which was most likely cited because of the Pope's concession of the definition of death to the medical profession ("A Definition of Irreversible Coma," 1968).[1]

The proverbial elephant standing in the room, both for the Harvard committee and in subsequent debates about brain death, was the role played by organ transplantation interests in the decision to describe the patients in question as "dead." The committee was formed within a month after the first heart transplant operation, which was performed in December 1967 by the South African surgeon Christaan Barnard and became a tremendous media event proclaimed around the world. Historians and other writers have debated whether the rise of intensive care (and ensuing dilemmas of discontinuation of life support) or the imperatives of organ transplantation were most relevant to the work of the committee. Certainly both issues contributed to some extent; some of the committee's members, notably the neurologist Robert Schwab, had a long-standing interest in the use of EEG as a prognostic aid in ventilator-dependent victims of coma, predating the transplantation controversy (Belkin, 2003). At the same time, the committee included famed transplant surgeon Joseph Murray, and its drafts and memos contain numerous references to the relevance of its work for organ donation. Its decision to reduce the time period for diagnostic confirmation from 72 to 24 hours was intended to facilitate organ procurement, as was the decision to make the EEG confirmatory rather than mandatory. Finally, the decision to describe this class of patients as "dead" rather than existing in "irreversible coma" clarified their suitability as organ donors (Giacomini, 1968).[2]

Still, it is important to emphasize that the Harvard committee defined brain death, quite conservatively, as absence of functioning of the whole brain (including the brainstem) rather than just the higher brain. This became a topic of considerable debate over the next decade. Some writers argued that any condition involving permanent loss of awareness could be regarded as brain death, because consciousness represented the essence of being human. Although Beecher himself at times seems to have been attracted by this argument (Pernick, 1988, pp. 3–33), it never gained widespread acceptance. State laws passed in the 1970s retained the Harvard "whole brain" definition of death. The 1981 Presidential Commission report, *Defining Death,* firmly rejected "higher brain" in favor of "whole brain" criteria, emphasizing the loss of integrative brainstem reflexes as well as consciousness (President's Commission, 1981).

Telling the story in terms of commissions and definitions belies the truth that the actual practice of defining brain death in the 1970s and 80s was less than straightforward. The role of the "confirmatory" EEG was particularly disputed. A lively transatlantic exchange followed a British 1980 television broadcast ("Transplants: Are the Donors Dead?") on the program *Panorama*, which alleged that U. K. physicians were misdiagnosing some patients as dead by relying solely on neurological examination of brainstem reflexes. American physicians, the program suggested, were much more inclined to confirm brain death with EEG and other ancillary testing. British neurologists responded somewhat testily that the greater American reliance on confirmatory diagnostic technology compensated for shoddy clinical examination skills. Moreover, brain death was conceptualized as "brainstem death" in the United Kingdom and therefore did not require demonstration of the absence of functions involving the cerebral hemispheres, such as those demonstrated on EEG. Eventually, the two countries' criteria converged, returning to the Harvard committee's original emphasis on brainstem examination (Pallis, 1983; Wijdicks, 2012). The controversy in the early 1980s on whether a flat EEG should be required to pronounce brain death nonetheless offers an interesting parallel to today's controversy regarding the use of neuroimaging to identify minimal consciousness.

VEGETATIVE STATE

Another major territory in the geography of unconsciousness was given a name in the 1970s. In 1972, neurosurgeon Bryan Jennett and neurologist Fred Plum proposed a new term, persistent vegetative state (PVS), to describe patients who began to open their eyes again, after 2 or 3 weeks in the sleep-like state of deep coma, yet showed no signs of awareness or responsiveness. These patients manifested a variety of behaviors that might appear purposeful to a layperson (e.g., roving eye movements, grasping of objects, scratching toward a noxious stimulus, smiling) but actually reflect the persistence of brainstem reflexes. The authors concluded with a memorable understatement, noting that their newly coined term was "likely to be discussed widely outside of the profession" (Jennett & Plum, 1972).

On April 15, 1975, a 21-year-old woman named Karen Ann Quinlan was admitted to a New Jersey hospital in a state of coma following intoxication from alcohol and diazepam ingestion. She was intubated and placed on a ventilator and fed by means of a nasogastric tube. As Karen's parents visited daily over the next several weeks, they watched her emerge into a PVS, although her neurologist did not use the term. She began to open her eyes and move her arms and legs, but without any signs of responsiveness. As they became convinced that Karen's chance of recovery was vanishingly small, the Quinlans decided to request that their daughter be taken off the ventilator. Their priest agreed, citing Pope Pius' statement allowing the withdrawal of "extraordinary" life support in patients who would otherwise be dying. Karen's physicians refused, however, claiming that removing the ventilator would directly violate their Hippocratic professional duty to help and not harm the patient, not to mention rendering the medical team liable for homicide under New Jersey law. On September 12, the Quinlans began legal proceedings that would eventually lead to the New Jersey Supreme Court. Widely publicized by the media, Karen's name became a household word. The New Jersey Supreme Court finally ruled in the Quinlans' favor on March 31, 1976. Notably, the court justified its decision not in terms of freedom of religion, as the Quinlans' legal team had requested, but the right to privacy, which included a right to refuse medical intervention. It turned to Karen's family as most likely to represent her interests and suggested that analogous situations be addressed by physicians and families with the consultation of hospital ethics committees.[3]

The Quinlan case launched what became known as the "right-to-die" movement, a widespread popular campaign to assert the right of patients to discontinue live support if in PVS. Advocates adhered to the courts' reliance on the right to privacy by grounding the ethics of withdrawing life support on the principle of patient autonomy. Advanced directives known as "living wills" were widely promoted, although they never gained as much acceptance as their champions had hoped. In most cases, families or other proxies served to represent the interests of the patient, backed up in cases of disagreement by hospital ethics committees and the courts. These procedures were refined and approved in a major report by the Presidential Commission on bioethics, *Deciding to Forego Life-Sustaining Treatment* (President's Commission, 1983).

The definition of what kinds of medical intervention could be legally refused was gradually expanded over the course of the 1980s to include artificial feeding and hydration (Angell, 1994; Filene, 1998, pp. 96–183). The most important case was that of Nancy Cruzan, a 30-year-old woman in PVS whose parents requested discontinuation of her feeding tube. The state of Missouri refused their request in the absence of "clear and convincing" evidence of Nancy's own wishes, propelling the case to the U.S. Supreme Court. The Justices, in 1990, found that there was a constitutional basis (14th Amendment) supporting the right of all American citizens to refuse unwanted medical therapy, including artificial hydration and nutrition, and that this right was not extinguished by incompetence but could be exercised on a person's behalf by a surrogate who could provide evidence of what the patient would have wanted. At the same time, the Court recognized the authority of the states to set their own standards of evidence for reconstructing the wishes of an incompetent patient, thereby upholding Missouri's requirement that Nancy's wishes had to be substantiated by "clear and convincing evidence." It was only after the case returned to the Missouri courts and some of Nancy's friends came forth with new evidence that would meet the "clear and convincing" standard that her guardians were authorized to withdraw treatment (Annas, 1990).

When Nancy's feeding tube was disconnected, pro-life protestors camped outside the hospital and at one point went so far as to try to break into her room and force-feed her. Her final days provided a graphic demonstration of how the care of patients in PVS had become politicized. Although many bioethicists saw the fundamental issue in the Cruzan case as no different from that of Quinlan—the right to refuse treatment—many religious groups saw a profound difference between disconnecting a ventilator and withdrawing food and nutrition. The Quinlans, ironically given their celebrity within the right-to-die movement, were devout Catholics, and when Karen remained alive after her extubation, they had continued to care for her with the aid of a feeding tube until she finally succumbed to pneumonia in 1985.

As the decade of the 1990s progressed, public debates about withdrawal of support for patients in PVS became conflated with those regarding patients with physical or intellectual disabilities. Particularly inflammatory was the so-called Baby Doe controversy, in which an infant with Down syndrome was allowed to die after the parents refused surgery for esophageal atresia (a severe malformation in which the esophagus ends in a blind pouch rather than connecting to the stomach). Religious conservatives frequently used "slippery slope" language to suggest that the trajectory from Quinlan to Cruzan would eventually lead to passive euthanasia of other "lives not worth living" among the elderly and disabled. Was the removal of a feeding tube in an adult with PVS fundamentally different from denial of nutrition or life-saving surgery for a child with severe intellectual disability? Were secular bioethicists too ready to equate the value of human life with cognitive ability? Or was the ability to form relationships the most important, and if so, what if a family interpreted the smiling and spontaneous movements of a person with PVS as evidence of interaction? (Filene, 1998, pp. 96–124).

Medical professionals responded with empirical studies and guidelines seeking to clarify the definition and management of PVS. The most important of these in the United States was the 1994 report of a multi-society task force representing North American neurologists, neurosurgeons, and pediatricians (Multi-Society Task Force, 1994). Notably, the sole philosophical reference cited by the task force was William James' 1890 definition of consciousness as having two dimensions, wakefulness and awareness. The authors did acknowledge the conceptual uncertainty of assessing "awareness" in an unresponsive patient, as well as the possibility of misdiagnosing some kind of "locked-in state" in which a patient might theoretically retain consciousness but lack any means to demonstrate it. The report appealed to three lines of evidence that vegetative patients truly lack awareness: their stereotypical movements, positive emission tomography scans showing greatly reduced glucose metabolism compared with patients in a locked-in state, and neuropathological examinations showing a degree of degeneration generally thought to make awareness highly improbable.

A major thrust of the report was to provide a clear definition of when a vegetative state was "persistent" versus "permanent," recognizing that the latter represents a prognostic rather than a diagnostic statement. After reviewing a substantial body of clinical studies, the task force concluded that a vegetative state could be considered "permanent" (in the sense that recovery is either impossible or, at the very least, associated with severe disability) 12 months after a traumatic injury or 3 months after a hypoxic-ischemic insult. This effort to define "permanent" should be seen in the context of a growing ethical problem in the 1990s that was very different from that posed by the Quinlan and Cruzan cases: What should be done when a family insists on continued treatment of a patient in PVS whose situation is deemed hopeless by medical staff? Some ethicists argued that life support in such a situation could be discontinued if judged to be medically futile. The courts rejected such reasoning, notably in the 1991 case of Helen Wanglie (a woman in PVS whose husband requested continued support over the objections of the medical staff) and that of Baby K (who was kept on a ventilator at the request of her mother despite anencephaly, a state seen as analogous to PVS from the standpoint of futility). In the absence of clearly expressed advance directives, surrogates continued to have the greatest say concerning the treatment of patients in vegetative states (Angell, 1994; Post, 1995).

The other type of ethical problem that can take place in PVS occurs when potential surrogates disagree among themselves. A spectacular example of what can go wrong in such circumstances was the case of Terri Schiavo. On February 25, 1990, Ms. Schiavo collapsed in the hallway of her apartment at age 26 from a cardiac arrest thought to be caused by an electrolyte disturbance later attributed to an eating disorder. She remained profoundly incapacitated, requiring percutaneous insertion of a feeding tube, and by the end of the year she was declared to be in PVS. Her husband, Michael, and her parents at first cooperated in trying a variety of aggressive therapeutic interventions throughout the decade, but by the mid-1990s, Michael had come to believe that his wife's condition would never improve and that she would not want to be kept alive in such a state. He failed to come to agreement with Terri's parents, however, and petitioned the guardianship court to function as a proxy and make is own recommendations. Thus began a roller coaster of court hearings extending from 2001 to 2005 to decide the Ms. Schiavo's fate, all widely reported in the media. The Schiavos posted selective video footage that they claimed showed evidence of their daughter's responsiveness. Artificial nutrition and hydration were discontinued and then reinstituted twice, the second time after the intervention of the Florida governor and state legislature, and finally removed at court order in 2005 despite the attempted intervention of the U.S. Congress. Terri Schiavo ultimately died on March 31, 2005 (Perry, Churchill, & Kirshner, 2005).

The Schiavo affair illustrates how politically charged the issue of withdrawing artificial nutrition and hydration from a person in PVS remained more than 20 years after the Quinlan case. Articles in the medical, bioethical, and legal literature were overwhelmingly critical of how media outlets covered the story (Annas, 2005; Quill, 2005). Popular accounts tended to reduce the complexities of the story to the level of dysfunctional family dynamics and culture war politics. Little coverage was given to the exhaustive court testimony provided by neurologists attesting to Ms. Schiavo's condition or by close friends attesting to her preferences. More troublingly, the affair revealed widespread lack of understanding of PVS. Many people simply were unaware that patients in a vegetative state have open eyes and may appear to smile or move (although not purposefully). The Schiavos thus had great success influencing public opinion by circulating an edited videotape allegedly showing their daughter to be capable of basic interaction. Among many other aspects of the tragedy, as some commentators noted, an opportunity for public education was lost (Bernat, 2008).

In the end, there can be little if any doubt that Terri Schiavo truly was in a vegetative state so profound that it could reasonably be characterized as permanent. Every neurologist appointed by the court to examine her found no signs of awareness or responsiveness. Ms. Schiavo's brain on autopsy was grossly atrophied, weighing only 615 g, less than half of that expected for an adult her age.

Nonetheless, what was true for Terri Schiavo was not necessarily true for all persons in vegetative states. Even as the court proceeded to determine her status, neurologists were once again redrawing the map of the disorders of consciousness.

MINIMALLY CONSCIOUS STATE

Over the course of the 1990s, a series of articles demonstrated that a disturbingly high proportion of patients admitted to rehabilitation units with the diagnosis of vegetative state (up to 40% in one study) in fact had minimal signs of awareness on later examination (Andrews, Murphy, Munday, & Littlewood, 1996; Childs, Mercer, & Childs, 1993; Wilson, Harpur, Watson, & Morrow, 2002). This does not necessarily mean that all such patients were misdiagnosed after their initial injury or insult; some may have gradually regained a limited degree of responsiveness. These behaviors may have been subtle and intermittent and could well have been missed on a single neurological examination. Quite possibly, the objections of family members may have been laid aside as amounting to denial or unrealistic hope.

In 2002, this borderline state of awareness was given a name: the minimally conscious state (MCS). The new term was coined by a multidisciplinary task force, the Aspen Neurobehavioral Conference Workgroup, which represented fields ranging from bioethics and neurology to physiatry and neuropsychology (Giacino et al., 2002). The diagnosis required identification of "limited but clearly discernible evidence of self or environmental awareness" on a reproducible or sustained basis, such as purposeful movement, responsive vocalizations or gestures, or appropriate smiling. The workgroup acknowledged that the line between this MCS and higher states of consciousness was necessarily arbitrary, suggesting that the demarcation be defined by consistent demonstration of functional interactive communication or the use of at least two objects. The report offered only limited evidence regarding prognosis but concluded by emphasizing the imperative of trying to establish functional communication in such persons.

One of the great challenges in thinking about MCS goes back to the basic dilemma of understanding consciousness: We may be able to describe what it means functionally, but we ultimately cannot experience the state of consciousness of another person. Therefore, the meaning of MCS can be understood in radically different ways. Some people may respond with renewed efforts to make contact with the person presumably bound by the constraints of her own body, calling on innovative therapies and communication strategies, both evidence-based and not. Others may imagine the prospect of facing life sustained by a feeding tube as still more unbearable for a person who is aware rather than fully unconscious.

Two patients' stories captured the opposite poles of these hopes and fears. On the side of hope was the widely reported "miracle awakening" of Terry Wallis, a 39-year-old man from Arkansas who made front-page headlines in 2003 when he began to speak 19 years after having received a diagnosis of vegetative state following a motor vehicle accident. In fact, his family had begun to see signs of awareness and interaction about a year after his injury, but the facility's staff dismissed their reports as impossible for a person in a vegetative state (Fins, Schiff, & Foley, 2007). There is no counterpart story for vegetative state or MCS describing the opposite scenario, how it might feel to possess awareness but be unable to communicate. The closest metaphor for this kind of existence might be the rare locked-in syndrome, exemplified by a person who is fully aware yet unable to interact with other people due to complete or near-complete paralysis (classically caused by a stroke of the brainstem). In 1997, the French editor-in-chief of *Elle* magazine, Jean Dominique Bauby, experienced such a state following a severe stroke that led to almost complete paralysis. Capable of communicating only by blinking his left eyelid in response to the presentation of alphabet blocks, Bauby was able to "dictate" a powerful memoir, *The Diving Bell and the Butterfly* (Bauby, 1997). Bauby's narrative conveys courage and the human drive to find meaning through life's most trying circumstances. A different kind of story was that of Tony Nicklinson, who described his locked-in syndrome occurring after a massive stroke in 2005 as a "living nightmare." He died in 2012 after unsuccessfully petitioning British Courts to authorize his doctors to end his life (Kmietowicz, 2012).

The point of mentioning locked-in syndrome is not that it is part of the continuum of minimal consciousness; patients in this state have clear and consistent awareness, in contrast to the fluctuating and intermittent state of awareness that seems to characterize MCS. But the locked-in syndrome has functioned as a metaphor for what a person in such a state might experience. It may be the modern counterpart of the fear of being buried alive.

NEW STUDIES, OLD HOPES AND FEARS

This history sets the stage for the most recent act in this story: the publication of two studies, in 2006 and 2010, that brought neuroimaging to bear on understanding MCS in a most dramatic way. The first was a case report published in *Science* by Adrian Owens and colleagues at Cambridge and Liege (Owen et al., 2006). It described the results of functional magnetic resonance imaging (fMRI) applied to a 23-year-old woman who was diagnosed as being in a vegetative stage 5 months after a traumatic brain injury resulting from a motor vehicle accident. The investigator instructed the subject to imagine two distinct tasks while under fMRI examination: playing a game of tennis and visiting each room of her house. The two activities were associated with strikingly different fMRI patterns involving different regions of the brain, indicating that this "vegetative" patient still had some ability to understand and respond to verbal commands.

In 2010, Martin Monti and the Cambridge/Liege teams published a still more remarkable neuroimaging study of 54 patients in a vegetative state or in MCS (Monti et al., 2010). Five patients were able to modulate their brain activity on fMRI in response to verbal imaging questions similar to those described by

Owens. In one patient, who had been considered to be in a vegetative state on his most recent neurological examination, the investigators were able to establish a channel of communication using the fMRI. They obtained consistent answers to five of six "yes or no" questions by asking the patient to use motor imagery for one response and spatial imagery for the other. Neurological reexamination of this patient confirmed subtle and fluctuating signs of awareness (suggesting MCS rather than vegetative state), but meaningful communication at the bedside was never established despite renewed efforts. It appeared that, at least in rare instances, it might be possible to communicate by fMRI with a patient in MCS even if no other technique has been successful.

These two studies raise many new questions for clinical medicine. Can fMRI identify a subgroup of patient who do not fit criteria for either vegetative state or MCS and who have evidence of awareness and responsiveness only on neuroimaging? What role should neuroimaging have in the routine evaluation of such patients? Can neuroimaging furnish a means to communicate with patients in an apparent vegetative state or MCS? How reliable and consistent are patients' answers obtained by fMRI? Can they be used to make patients more comfortable and improve their quality of life? Will there be a day when neuroimaging-based communication plays a role even in end-of-life decisions?

Rigorous clinical research will be required to answer these questions. It can be expected, however, that MCS will generate heated and polarized discussion in the framework of popular culture. Some will see it as part of the great narrative of medical technology's ever-increasing ability to challenge the inevitability of death, a story that we have followed from the Enlightenment's infatuation over resuscitation of drowning victims through the rise of modern intensive care. For others, it will trigger fears of abandonment and isolation, of ending one's life locked in a functionless body and unable to reach out to others—our modern equivalent of the old fear of premature burial. Such hopes and fears mirror our conflicted attitudes toward medical technology and dying.

NOTES

1. The Roman Catholic Church has more recently changed its position and at present considers the acceptance of brain death as within its purview, although it continues to support the concept; see Edward J. Furton (2002), "Brain Death, the Soul, and Organic Life," *National Catholics Bioethics Quarterly, 2,* 455–470.
2. With the notable exception of Gary Belkin, most historians and bioethicists have emphasized the role of organ transplantation over end-of-life care in the Harvard definition, most notably Martin S. Pernick, "Brain Death in a Cultural Context: The Reconstruction of Death, 1967-1981" (in *The Definition of Death: Contemporary Controversies*; Stuart J. Younger, Robert M. Arnold, & Renie Schapiro, Eds.;

Baltimore, Johns Hopkins University Press, 1999, pp. 3–33), and David J. Rothman, *Strangers at the Bedside: A History of How Law and Bioethics Transformed Medical Decision Making* (New York, Basic Books, 1991, pp. 148–167).

3. The most thorough account of the Quinlan case is that of Peter G. Filene (1998, pp. 11–95).

REFERENCES

Andrews, K., Murphy, L., Munday, R., & Littlewood, C. (1996). Misdiagnosis of the vegetative state: Retrospective study in a rehabilitation unit. *British Medical Journal, 313*, 13–16.

Angell, M. (1994). After Quinlan: The dilemma of the persistent vegetative state. *New England Journal of Medicine, 330*, 1524–1525.

Annas G. (1990). Nancy Cruzan and the right to die. *New England Journal of Medicine, 323*, 670–673.

Annas, G. J. (2005). "Culture of life" politics at the bedside: The case of Terri Schiavo. *New England Journal of Medicine, 353*, 1710–1715.

Bauby, J.-D. (1997). *The diving bell and the butterfly.* (J. Leggart, Trans.). New York: Vintage Press.

Bauer, G. (1987). Coma and brain death. In E. Niedermayer & F. Lopez de Silva (Eds.), *Electroencephalography.* Baltimore: Urban and Schwartzberg.

Belkin, G. S. (2003). Brain death and the historical understanding of bioethics. *Journal of the History of Medicine and Allied Sciences, 58*, 343–346.

Bernat, J. (2008). Theresa Schiavo's tragedy and ours, too. *Neurology, 71*, 964–965.

Childs, N. L., Mercer, W., N., & Childs, H. W. (1993). Accuracy of diagnosis of persistent vegetative state. *Neurology, 43*, 1465–1467.

A definition of irreversible coma: Report of the Ad Hoc Committee of the Harvard Medical School to Examine the Definition of Brain Death. (1968). *Journal of the American Medical Association, 205*(6), 337–340.

Diem, S. J., Lantos, J. D., & Tulsky, J. D. (1996). Cardiopulmonary resuscitation on television: Miracles and misinformation. *New England Journal of Medicine, 334*, 1578–1582.

Eisenberg, M. S. (1997). Life in the balance: Emergency medicine and the quest to reverse sudden death. New York: Oxford University Press.

Filene, P. G. (1998). In the arms of others: A cultural history of the right-to-die in America. Chicago, Ivan R. Dee.

Fins, J. J., Schiff, N. D., & Foley, K. M. (2007). Late recovery from the minimally conscious state: Ethical and policy implications. *Neurology, 68*, 304–307.

Giacino, J. T., Aschwal, S., Childs, N., Cranford, R., Jennett, B., Katz, D. I., . . . & Zasler, N. D., (2002). The minimally conscious state: Definition and diagnostic criteria. *Neurology, 58*, 349–353.

Giacomini, M. (1997). A change of heart and change of mind? Technology and the redefinition of death in 1968. *Social Science and Medicine, 44*, 1465–1482.

Hawes, W. (1780). *An address to the public on premature death and premature interment.* London: For the author.

Jennett, B., & Plum, F. (1972). Persistent vegetative state after brain damage: A syndrome in search of a name. *Lancet, 1*, 734–737.

Jonsen, A. R. (2008). Encephaloethics: A history of the ethics of the brain. *American Journal of Bioethics, 8*, 37–42.

Kmietowicz, Z. (2012). Man with locked-in syndrome who fought for doctors to end his life has died. *British Medical Journal, 345*, e5729.

Koehler, P. J., & Wijdicks, E. F. M. (2008). Historical study of coma: Looking back through medical and neurological texts. *Brain, 131*(Pt 3), 877–889.

Lathrop, J. (1787). *A discourse before the Humane Society in Boston*. Boston: E. Russell.

Markel, H. (1994). The genesis of the iron lung: Philip Drinker, Charles F. McKhan, James L. Wilson, and early attempts at administering artificial respiration to patients with poliomyelitis. *Archives of Pediatrics and Adolescent Medicine, 148*, 1174–1180.

Mollaret, P., Bertrand, I., & Mollaret, H. (1959). Coma depassé et necroses nerveuses centrales massives. *Revue Neurologique, 101*, 116–139.

Monti, M. M., Vanhaudenhuyse, A., Coleman, M., Boly, M., Pickard, J. D., Tshibanda, L., . . . & Laureys, S. (2010). Willful modulation of brain activity in disorders of consciousness. *New England Journal of Medicine, 362*, 579–589.

Moore, W. (2005). The knife man: Blood, body snatching, and the birth of modern surgery. New York: Broadway Books.

Moruzzi, G., & Magoun, H. W. (1949). Brain stem reticular formation and activation of the EEG. *EEG Clinics in Neurophysiology, 1*, 455–473.

The Multi-Society Task Force on PVS. (1994). Medical aspects of the persistent vegetative state. *New England Journal of Medicine, 330*, 1499–1508, 1572–1579.

Owen, A. M., Coleman, M. R., Boly, M., Davis, M. H., Laureys, S., & Pickard, J. D. (2006). Detecting awareness in the vegetative state. *Science, 313*, 1402.

Pallis, C. (1983). ABC of brain death: The arguments about the EEG. *British Medical Journal, 286*, 284–287.

Pernick, M. S. (1988). Back from the grave: Recurring controversies over defining and diagnosing death in history. In R. M. Zaner (Ed.), *Death: Beyond whole-brain criteria*. Dordrecht: Kluwer Academic Publishers.

Perry, J. E., Churchill, L. R., & Kirshner, H. S. (2005). The Terri Schiavo case: Ethical, legal, and medical perspectives. *Annals of Internal Medicine, 143*, 744–748.

Post, S. G. (1995). Baby K: Medical futility and the free exercise of religion. *Journal of Law, Medicine, and Ethics, 23*, 20–26.

Powner, D. J., Ackerman, B. M., & Grenvik, A. (1996). Medical diagnosis of death in adults: Historical contributions to current controversies. *Lancet, 348*, 1219–1223.

President's Commission for the Study of Ethical Problems in Medicine and Biomedical and Behavioral Research. (1981). *Defining death: A report on the medical, legal, and ethical issues in the determination of death*. Washington, DC: U.S. Government Printing Office.

President's Commission for the Study of Ethical Problems in Medicine and Biomedical and Behavioral Research. (1983). *Deciding to forego life-sustaining treatment: Ethical, medical, and legal issues in treatment decisions*. Washington, DC, U.S. Government Printing Office.

Quill, T. E. (2005). Terri Schiavo: A tragedy compounded. *New England Journal of Medicine, 352*, 1630–1633.

Reiser, S. J. (1992). The intensive care unit: The unfolding and ambiguities of survival therapy. *International Journal of Technology Assessment in Health Care, 8*, 382–394.

Snider, G. L. (1989). Historical perspective on mechanical ventilation: From simple life support system to ethical dilemma. *American Review of Respiratory Disease, 140*, S2–S7.

Struve, C. A. (1803). *A practical essay on the art of recovering suspended animation.* Albany, NY: Whiting, Backus, and Whiting.

Watkins, R. (1651). *Newes from the dead, or, A true and exact narration of the miraculous deliverance of Anne Greene.* Oxford: Lichfield, for Thomas Robinson. Retrieved from Early English Books On Line, http://eebo.chadwyck.com/search/full_rec?-SOURCE=pgimages.cfg&ACTION=ByID&ID=V151870.

Wijdicks, E. F. M. (2012). The transatlantic divide over brain death determination and he debate. *Brain, 135*, 1321–1331.

Wilson, F. C., Harpur, J., Watson, T., & Morrow, J. I. (2002). Vegetative state and minimally responsive patients: Regional survey, long-term case outcomes and service recommendations. *NeuroRehabilitiation, 17*, 231–236.

Wolstenholm, G. E. W., & O'Connor, M., (Eds.) (1966). *CIBA symposium: Ethics in medical progress with special reference to transplantation.* London: Churchill.

4

Consciousness and Death

The Whole-Brain Formulation of Death

JAMES L. BERNAT

Although an analysis of the definition of death is an unusual topic in a symposium on consciousness, it fits when the subject is viewed within its historical context. In the mid-20th century, attention to the definition of death was stimulated by reports of severely brain-damaged patients with total unconsciousness who were maintained on mechanical ventilators—leading to the recognition of a previously undescribed profound state dubbed *coma dépassé* (beyond coma) by French neurologists (Mollaret & Goulon, 1959). These observations led to the provocative claim that patients were dead when their brain damage was so severe and widespread that it abolished all clinically measureable brain functions. Thus, the state that later would be called "brain death" initially arose from descriptions of profound unconsciousness. But how did these cases lead to changes in the definition of death?

Before the 1950s, death always was a unitary phenomenon because of the interdependence of respiratory, circulatory, and brain functions. When one of these "vital" functions ceased as a consequence of illness or injury, the others stopped within minutes because no technological means existed to support the other vital functions. Therefore, when someone suffered a cardiac arrest, that person always also suffered respiratory arrest, and vice versa. Similarly, when brainstem functions failed as a consequence of brain injury or illness and spontaneous respiration ceased as a result, cessation of circulation followed rapidly and inevitably. Because of the interdependence of the vital functions, no one had to consider whether a patient was dead or alive who had completely lost all brain functions while maintaining circulatory and respiratory functions, because such cases were technologically impossible.

The implementation of tracheal positive-pressure ventilation (TPPV) into medical practice in the 1950s permanently altered the interdependence of cessation of vital functions. For the first time, patients who had suffered irreversible cessation of all brain functions (including spontaneous respiration) could have their absent respiratory function successfully replaced by TPPV, thereby preventing the otherwise inevitable immediate circulatory arrest. This mechanical or "artificial" maintenance of respiration and circulation permitted the continued functioning of bodily organs, other than the brain, at least temporarily. The technological breakthrough of TPPV was rightly hailed as a marvelous advance in saving the lives of otherwise healthy patients with temporary respiratory failure, but, by supporting the respiration of patients who had lost all brain functions, it also spelled the end to human death as a unitary phenomenon.

Once death was no longer a unitary phenomenon, the essential question needed to be answered: Were patients alive or dead whose brain functions all had ceased irreversibly but whose absent spontaneous respiratory and circulatory functions were mechanically supported? Their precise status became ambiguous because they shared characteristics of both living and dead persons. Like living patients, their autonomously beating hearts circulated blood oxygenated by the mechanical ventilator, allowing their organs (except the brain) to remain functioning. But, like dead patients, they could not breathe or move at all, showed no behavioral or reflex response to any stimulus, and lacked internal homeostatic processes.

The pioneering physicians who first described these unresponsive patients intuited that they differed in an essential way from all other patients in coma who had ever been examined. In a landmark article, a committee of Harvard Medical School physicians and scholars coined the misleading term "brain death" for these patients, provided diagnostic criteria for physicians to apply, and asserted that these patients were dead because of the irreversible cessation of their brain functions, despite the continued presence of their circulation, heartbeat, and visceral organ functioning. (A Definition of Irreversible Coma, 1968).

The Harvard Medical School committee report is often credited with catalyzing a transformative social change by positing a new definition of death. A more accurate depiction is that TPPV support permitting a diagnosis of brain death did not change the definition of human death but created an example of a previously unanticipated biological state, showing that our previous, simple, unitary definition of death had become inadequate. Scholars who attempted to answer the question of whether the brain-dead patient was dead or alive first had to more precisely define death by making explicit the meaning that is implicit in our ordinary use of the word "death" and had been rendered ambiguous by TPPV.

Starting in the 1970s, scholars began to provide refined definitions and criteria of death. Noteworthy in this regard were the efforts of the following:

- The Task Force on Death and Dying of the Institute of Society, Ethics, and the Life Sciences (later renamed the Hastings Center) in 1972, which showed the consistency of the emerging concept of brain death with both secular philosophical analysis and religious teachings
- Capron and Kass (1972), who first stratified levels of analysis and emphasized the need for society to agree on a concept of death before physicians could devise tests by which it could be measured
- Korein (1978), who, using thermodynamic theory, emphasized that the brain is the critical system of the organism whose destruction is death
- The President's Commission for the Study of Ethical Problems in Medicine and Biomedical and Behavioral Research (1981), which provided a rigorous explanation of why brain-dead patients are dead and offered a model uniform statute of death incorporating a brain criterion
- Pallis (1983), who proposed the brainstem as the essential component of "the brain as a whole" whose irreversible absence of functions was both a necessary and sufficient condition for death.

A BIOPHILOSOPHICAL ANALYSIS OF DEATH

My colleagues, Bernard Gert and Charles Culver, and I contributed to this debate by offering a rigorous biophilosophical analytical method, proceeding from the conceptual to the tangible and measurable (Bernat, Culver, & Gert, 1981; 1982) that since has been accepted by other scholars. For example, the President's Commission cited our analysis in their classic work, *Defining Death* (1981), as part of their discussion of the conceptual foundation for the equivalency of brain death and human death. Even many scholars who disagreed with one or more elements of our proposed definition and criterion of death accepted our analytical approach. For example, Alan Shewmon, a scholar who completely rejects the brain criterion of death in favor of the circulatory criterion, described our analytical method as "virtually universally accepted" (Shewmon, 2010). Our sequential, systematic analytical method provides a valuable tool because it proceeds in a logical order that permits one to pinpoint areas of scholarly disagreement and clarify the reasons for it. Here, I present an outline and biophilosophical justification of our analysis. Elsewhere, I have responded to critics of each level of our analysis (Bernat, 2013b).

An interesting irony is the striking contrast in attitudes between scholars and medical practitioners. The controversies over the definition and criterion of death that have raged within the academy for the past 40 years are largely absent among medical practitioners. The medical tests for determining brain-based and circulation-based death are so well accepted by the medical establishment that current controversies are restricted to technical details. Similarly, despite minor differences among death statutes in different jurisdictions, the legal standards for physicians to declare death are more or less uniform (Bernat, 2013a). Thus, the claims of some prominent brain death critics over the past few decades announcing "the death of whole-brain death" (Veatch, 2005) or the "impending collapse of the whole-brain definition of death" (Veatch, 1993) or asking whether "brain death can be resuscitated" (Shewmon, 2009) have proved to be inaccurate in medical practice.

The Bernat-Culver-Gert analytical method comprises four sequential components: paradigm, definition, criterion, and tests (Bernat, 2002). The *paradigm* is a set of preconditions and assumptions that frame the argument by identifying its nature, clarifying the class of phenomena to which it belongs, and demarcating its conceptual boundaries (Bernat, 2002). Accepting the paradigm conditions is a prerequisite for coherent discussion so that scholars can discuss the same class of phenomenon. Identifying the *definition* of death is the philosophical task of making explicit the meaning implicit in our ordinary use of the common and nontechnical term "death" that we all use correctly. Identifying the *criterion* of death is the philosophical and medical task of determining the measurable general condition that shows that the definition has been fulfilled by being both necessary and sufficient for death. Devising *tests* of death is the medical-scientific task whose goal is to demonstrate that the criterion of death has been fulfilled with no false-positive and minimal false-negative determinations.

THE PARADIGM OF DEATH

Our paradigm of death has seven conditions:

1. The word "death" is a common, nontechnical word that we all use correctly to refer to the cessation of the life of a human being or another higher vertebrate species. The philosophical task of defining death seeks not to redefine it by contriving a new meaning but rather to make explicit the implicit meaning of death underlying

our consensual usage of the word, which has become ambiguous as a result of technological advances such as TPPV. Scholars should neither redefine death away from its ordinary meaning to achieve an ideologically desired end nor overanalyze it to such a metaphysical level of abstraction that it is rendered devoid of its ordinary meaning.

2. Death is fundamentally a biological phenomenon. Because life is fundamentally a biological phenomenon, its cessation also must be. This condition does not denigrate the value of cultural and religious practices surrounding death and dying, nor does it deny societies the authority to establish laws regulating the determination and time of death. Because death is an immutable biological fact and is not socially contrived, our concern is restricted to the ontology of death and not its normative aspects.
3. We restrict our analysis to the death of higher vertebrate species for which death is univocal—that is, we mean the same concept of death when we say our cousin died as we do when we say our dog died. Simpler animals, plants, and parts of organisms, such as cells or organs, also can die, but our focus here is on the death of the higher animal organism.
4. The term "death" can be applied directly and categorically only to organisms. All living organisms must die, and only living organisms can die. When we say "a person died," we are referring to the death of the formerly living human organism that embodied the person, not to a human organism that remains alive but ceases to have the attributes of a person. Personhood is a psychosocial, religious, moral, and legal construct that may be lost in some cases of severe brain damage but cannot die, except metaphorically.
5. A higher organism can reside in only one of two states, alive or dead. No organism can be in both states or in neither. Death does not occur in degrees of severity, but parts of organisms can die while other parts remain alive, often because of technological support. Yet, we currently lack the technical ability always to accurately identify an organism's state, and, at times, we may determine it with confidence only in retrospect. Alive and dead therefore are mutually exclusive (non-overlapping) and jointly exhaustive (no other) biological states.
6. Death is conceptualized most accurately as an event and not a process. If there are only two exclusive and non-overlapping underlying states of an organism, the transition from one state to the other, at least in theory, must be sudden, discontinuous, and instantaneous, because of the absence of an intervening state. For technical reasons, the event of death may be determinable with

confidence only in retrospect. As my colleagues and I observed more than 30 years ago, death is best conceptualized not as a process but as the event separating the biological sequential processes of dying during life and bodily disintegration after death (Bernat, Culver, & Gert, 1981).

7. Death is irreversible. If the event of death were reversible, it would not be death but rather incipient dying that was interrupted and reversed. Reports of alleged return from the dead are accounts of "near-death" experiences occurring while alive and are thought to arise from disturbed brain physiology in critical illness (Parnia, Waller, Yeates, & Fenwick, 2001). Accounts of so-called afterlife in heaven or hell and of the continued existence after death of the disembodied soul are theories or religious beliefs.

THE DEFINITION OF DEATH

Because death is a biological event, defining death requires an understanding of fundamental biological concepts—most relevantly, the organism as a whole. The "organism as a whole" concept emphasizes the distinction between the life-state of the organism itself and the life-state of its component parts. A definition of death addresses the level of the life-state of the organism, not of its component parts. Technological support may allow the curious situation in which component parts remain alive despite the death of the organism.

An organism is composed of hierarchically arrayed interdependent units, each of which is alive but none of which alone constitutes the organism. The organism's cells, tissues, organs, and organ systems are organized in functional groups, producing hierarchies of functions. The interrelationships of the many hierarchies of functional units create an integrated, coordinated, unified whole. That whole is the organism itself, the highest and most complex life form that is alive as a result of the functioning of its living component subsystems.

Functioning components of an organism generate unique phenomena known as *emergent functions*. An emergent function is a function of a whole entity that is neither present within nor can be reduced to any of its component parts (Mahner & Bunge, 1997). Such a function is called "emergent" because, given the normal coordinated presence of bodily components in an operational unit, the new function emerges spontaneously. Each emergent function is a more complex behavior that any of the functions of its component subunits. Given our current understanding of biology and mathematical modeling, emergent functions cannot be accurately predicted or easily understood solely by studying the component subunits, their interrelationships, and their functions (Clayton & Kauffman, 2006). The most inscrutable example of an emergent function is that of human conscious awareness, an exquisite but ineffable phenomenon that is widely believed to emerge spontaneously from the integrated functioning of multiple distributed parallel hierarchical networks of brain neurons (Koch, 2004).

The life of the cellular, tissue, or organ components, while often necessary for the life of the organism, is not equivalent to the life of the organism. Because the life of its component parts is not equivalent to the life of an organism, an organism can die even if some of its component parts remain alive as a consequence of technological support. The key to understanding the definition of death centers on the distinction between the life of an organism and the life of its component parts.

In our technological era, the most accurate definition of death is *the irreversible cessation of functioning of the organism as a whole.* The concept of "organism as a whole" was proposed by the early 20th century biologist, Jacques Loeb, in a classic monograph of that name (Loeb, 1916). The "organism as a whole" refers not to the whole organism (the sum of its component parts) but rather to the emergent functions of the organism that are the consequence of the normal operation of, but greater than, the mere sum of its component parts. Intrinsic to the concept of "organism as a whole" is that the interrelatedness of the component parts provides emergent functions that create the coherent unity of the organism.[1]

In our contemporary technological era in which skilled physicians with advanced technology can maintain the life of component parts of organisms outside of or inside the body, the continued life of the organism has been wrongly attributed to the continued life of its component parts. The essence of the death of a higher animal species is the irreversible cessation of the functioning of the organism as a whole. Once an organism has irreversibly lost its totality, completion, indivisibility, self-reference, and identity, it no longer functions as a whole and is dead.

THE CRITERION OF DEATH

The criterion of death is the measurable general condition, suitable for inclusion in a death statute, that shows that the definition has been fulfilled by being both necessary and sufficient for death. In published analyses of death, four principal choices for a criterion of death have been proposed, and each has been defended by scholars: the whole-brain, higher-brain, brainstem, and circulatory formulations. The first three are variants of the brain death concept, whereas the circulatory formulation rejects brain death and holds that a person is not dead until systemic circulation ceases irreversibly. An overwhelming majority of medical associations and jurisdictions have accepted the practice of brain death and accept the whole-brain criterion of death that my colleagues and I also accept. The brainstem formulation prevails in the United Kingdom, yet the tests for "brainstem death" there and for brain death elsewhere are almost identical (Pallis, 1983).

The higher-brain formulation posits that the human organism is dead when consciousness and cognition are irreversibly lost because it is these characteristics that make human existence unique (Veatch, 1993). The higher-brain formulation would declare dead those patients with diffuse cortical and thalamic neuronal loss (e.g., a patient in an irreversible vegetative state). Despite its popularity among some philosophers and lawyers over the last 40 years, no medical society or jurisdiction has adopted the higher-brain formulation as a determination of human death. Elsewhere I have analyzed the shortcomings of the higher-brain formulation, which have led to its consistent disregard by physicians and policymakers (Bernat, 1992; 1998). The rightful place of the higher-brain formulation is not as a criterion of death but in determining the loss of personhood, a condition that might lead to a decision to withhold further life-sustaining therapy in a severely brain-damaged living person.

The whole-brain criterion is necessary for death because the operations of the functions of the organism as a whole are distributed throughout the brain. The brainstem contains centers of respiration and circulation, as well as the reticular system that is necessary for wakefulness, a prerequisite for conscious awareness. The diencephalon contains centers for neuroendocrine and autonomic control and homeostasis, integration of sensory input and motor outputs, and conscious awareness. The cerebral hemispheres and thalamus are necessary for conscious awareness.

The irreversible cessation of brain functions serves as a criterion of death because it is a necessary and sufficient condition for the cessation of the organism as a whole. A brain-dead patient whose visceral organ functions are maintained only as a consequence of technological support has lost the functions of the organism as a whole and is only a living component part of a dead organism, analogous in type, although not in extent, to a technologically supported isolated living organ or limb. The irreversible loss of the functions of the brain that are responsible for the emergent functions of the organism as a whole indicates that the brain-dead patient is a mechanically supported, living component part of a human organism who has already died.

THE TESTS OF DEATH

Physicians may choose from two general sets of tests for death, depending on the clinical situation. In the overwhelming majority of circumstances in which

TPPV or cardiopulmonary resuscitation (CPR) is not provided or is unsuccessful, physicians determine death in the traditional way, using tests that show the permanent absence of circulation and breathing. Only in the rare and special circumstance in which a patient is undergoing TPPV or CPR (i.e., circulation and ventilation are supported) must the specific brain death tests be applied.

Although the circulatory-respiratory tests are distinct from the whole-brain tests, they are causally related. The circulatory-respiratory tests are valid because, once they are fulfilled, the brain is completely deprived of blood flow, and all brain neurons and other brain cells die from lack of oxygen and glucose. If brain perfusion is protected and brain function continues, the person is not dead. The whole-brain criterion therefore is the primary criterion of death. It can be shown to be fulfilled directly by the use of brain tests or indirectly by circulatory-respiratory tests.

Numerous batteries of brain death tests have been studied and validated since the Harvard Ad Hoc Committee report in 1968. Those in the United States and most parts of the world employ tests for the whole-brain criterion. Those in the United Kingdom and a few other countries use tests for the brainstem criterion. No test batteries in any country employ tests for the higher-brain criterion. The tests for the whole-brain and brainstem criteria are essentially identical except in the highly unusual case of a primary brainstem catastrophe (Ogata, Imakita, Yutani, Miyamoto, & Kikuchi, 1998); in that circumstance, it would be possible to be declared brain dead in the United Kingdom but not in the United States. However, this exception occurs so rarely in medical practice that most physicians have never seen such a case. In the United States, the currently accepted test battery for adults has been published by the American Academy of Neurology (Wijdicks, Varelas, Gronseth, & Greer, 2010); the currently accepted test battery for infants and children was published by a pediatric multi-society task force (Nakagawa et al., 2011).

Current medical controversies in brain death testing include whether two sequential examinations separated by a time interval should be required or a single examination is sufficient; whether ancillary tests showing the absence of intracranial blood flow are useful to confirm the clinical assessment; how to better standardize test batteries among institutions and countries; how long to wait to perform testing following the completion of protocols of therapeutic hypothermia after cardiac arrest to ensure irreversibility; the best way to respond to families who oppose brain death testing of their loved one for emotional or religious reasons; and how best to request organ donation (Bernat, 2013a).

Testing for death using the traditional circulatory-respiratory tests has become controversial recently in the context of organ donation after circulatory determination of death (DCDD), formerly known as non-heart-beating organ donation or organ donation after cardiac death. A "controlled" DCDD protocol may be conducted in the following circumstance: The lawful surrogate decision maker for a severely brain-damaged but not brain-dead patient in the intensive care unit who requires TPPV refuses further life-sustaining therapy on behalf of the patient, in light of the patient's poor prognosis and honoring the patient's prior stated wishes not to be treated in this situation, and requests that the patient serve as an organ donor after death. A "controlled" DCDD protocol coordinates the timing of the process of withdrawing life-sustaining treatment, which allows the patient to die, with the readiness of the surgical team to perform organ donation as rapidly as possible after the declaration of death (Bernat et al., 2006). An "uncontrolled" DCDD protocol, used in some European countries and experimentally in the United States, permits organ donation after death is declared in the setting of unexpected cardiopulmonary arrest with failure of resuscitation by CPR (Munjal et al., 2013).

Once the prospective "controlled" DCDD organ donor has had life-sustaining therapy discontinued, leading to stoppage of breathing and heartbeat (asystole), the principal controversy concerns how long physicians must wait to declare the patient dead. Although some observers claim that the patient is immediately dead once breathing ceases and the heart stops beating, physicians know that, because it might be possible to resuscitate the patient after the heart stops beating (despite the fact that CPR will not be performed in this circumstance) and because death is irreversible, the patient is not dead at the moment of cessation of vital functions.

An expert panel, recruited by the division of the U.S. Department of Health and Human Services that funds experimental protocols of organ donation, studied this question and offered guidelines for circulatory-respiratory tests of death for patients serving as organ donors in controlled DCDD programs. These guidelines also apply to determining death on circulatory grounds in non-donation circumstances, but in those cases, the precise timing of the death declaration usually is inconsequential. The panel pointed out that the justification for determining the moment of the death declaration turns on the distinction between "permanent" and "irreversible" cessation of circulation (Bernat et al., 2010). More recently, the panel issued related guidelines for death determination in emerging uncontrolled DCDD programs (Bernat et al., 2014).

A *permanent* cessation of a function means that the function *will not* be restored because it will not recover spontaneously and no medical attempts will be made to restore it. By contrast, an *irreversible* cessation of a function means that it *cannot* be restored by available technology even if medical attempts were made. Many death statutes stipulate the "irreversible" cessation of circulation and respiration, although medical standards for death declaration always have relied on the permanent cessation of circulation and respiration (Bernat, 2010a). Death determination in DCDD most starkly shows the noncongruence between the two conceptual approaches to death determination: The medical determination of death requires showing only the permanent cessation of circulation, whereas the biological-legal concept of death requires showing the irreversible cessation of circulation (Bernat, 2013d).

In the context of controlled DCDD, once the period has elapsed during which the asystolic heart could resume beating spontaneously (auto-resuscitation) and given that the do-not-resuscitate (DNR) order precludes resuscitative attempts, the patient can be declared dead once circulation and breathing have ceased permanently. As is true in death declaration for terminally ill patients with DNR orders who are expected to die and are not organ donors, once circulation and respiration have ceased permanently, it is unnecessary for physicians to delay the declaration of death for an additional 60 minutes or so until the cessation can be presumed to be irreversible (Bernat et al., 2010). Neither is it necessary to attempt to resuscitate the patient to prove that the cessation is irreversible. The permanent cessation of circulation standard in DCDD is implicit in all controlled DCDD protocols and was made explicit by the American Academy of Pediatrics (AAP) Committee on Bioethics in their recent statement on the ethics of DCDD (AAP, 2013).

Institutional protocols for controlled DCDD differ concerning the mandated duration of asystole before death can be declared. Part of the issue turns on an empirical question: What is the greatest duration after asystole in which auto-resuscitation has been reported? The most comprehensive study of auto-resuscitation showed that not a single case of auto-resuscitation to circulation has ever been reported after discontinuation of life-sustaining therapy (analogous to the process in controlled DCDD), although cases of auto-resuscitation to restored circulation have been reported after as long as 7 minutes of asystole following failed CPR (Hornby, Hornby, & Shemie, 2010). Therefore, auto-resuscitation to restored circulation is not a significant problem in controlled DCDD protocols but is a significant problem in uncontrolled DCDD protocols (Bernat, 2010b).

The AAP expert panel argued that the general standard for physicians applying the circulatory criterion of death should be the same in organ donation and non-donation circumstances; namely, certifying the permanent cessation of circulation and respiration. But because of the consequentiality of the determination of death in organ donation, the panel required physicians to perform a more precise assessment of the complete absence of circulation than simply palpating the patient's pulse or listening for heart sounds; they recommended the use of sensitive techniques to detect circulation, such as intra-arterial pressure monitoring, Doppler ultrasound arterial studies, or echocardiography to assess opening and closing of the aortic valve (Bernat et al., 2010).

RELIGIOUS VIEWS

Shortly after the introduction of the concept of brain death, a multidisciplinary task force of scholars empaneled by what now is called the Hastings Center studied the scientific, social, and legal issues of this newly described phenomenon. In an influential two-part article in the *Journal of the American Medical Association*, they argued that the emerging concept of brain death and its equivalence to human death was fully consistent with traditional Judeo-Christian teachings (Task Force, 1972). While this approach remains generally true, the breadth of these claims has been disputed, and religious authorities have written extensive analyses of this topic during the 4 decades since the Task Force report.

Within Christianity, Protestant denominations uniformly accept brain death. The magisterium of the Roman Catholic Church considered death determination to be a medical issue and not a matter of church doctrine until the 1990s, when the topic of brain death became the subject of intense debate by several Vatican pontifical academies and councils. Endorsement of the equivalence of brain death as human death by these councils in the 1990s led Pope John Paul to pronounce, in 2000, that brain death is fully consistent with Roman Catholic teachings and tradition (Furton, 2002). After the death of Pope John Paul, the diagnosis and signs of brain death were scrutinized by the Vatican Pontifical Academy of Sciences at the request of Pope Benedict, and this added further Roman Catholic ecclesiastical endorsement (Pontifical Academy of Sciences, 2007).

Within Judaism, the matter is more complex because, unlike the top-down rulings on religious doctrine in Roman Catholicism, Jewish religious doctrine is not issued by a central authority but accrues like case law as the result of learned discourse by rabbinic scholars who attempt to apply ancient Jewish law to contemporary problems. As might be expected, this type of scholarship leads to disagreements in interpretation of Talmudic sources that yield differences in acceptance. Currently, brain death is uniformly accepted among Reform and Conservative rabbinic scholars, but within Orthodox Judaism, there remains an active rabbinic debate. The strictest Orthodox rabbis, including the ultra-Orthodox, reject brain death and require circulatory and respiratory cessation for death to be declared (Rosner, 1999). By contrast, more biologically inclined Orthodox rabbis accept brain death (Halachic Organ Donor Society, 2012).

Within Islam, the situation is similarly complicated with varying degrees of acceptance of brain death by different influential and respected imams. As a result, although this approach is commonly practiced, it remains frequently accepted but non-uniform among Islamic nations (Padela, Arozullah, & Moosa, 2013). Hindu authorities in India accept brain death, which is practiced widely (Jain & Maheshawari, 1995). Following a cultural battle lasting for several decades and pitting traditional Shinto and Buddhist practices against Western medical developments, Japan now accepts the practice of brain death for organ donation under certain conditions (Lock, 1995).

A challenging situation arises in those cases in which physicians intend to declare brain death according to prevailing medical practice and law but the patient's religious convictions do not permit it. Successfully handling these cases requires respect for religious beliefs, knowledge of the relevant law, and compassion (Olick, Braun, & Potash, 2009). Scholars have debated the wisdom and permissible extent of providing variations or exceptions in public laws for declaring death before the allowed variations render the medical standard chaotic and unworkable (Miles, 1999).

In the United States, the states of New Jersey and New York have provided legal means for family members of patients whose religion does not accept brain death to forbid physicians from making the declaration and to require cessation of heartbeat and circulation for death to be declared. New Jersey enacted a statute containing such a religious exemption (Olick, 1991), whereas New York created a similar provision in its Department of Health administrative law (Beresford, 1999). The relevant portion of the New Jersey statute provides the following:

> The death of an individual shall not be declared upon the basis of neurological criteria . . . when such a declaration would violate the personal religious beliefs of moral convictions of that individual and when that fact has been communicated to, or should . . . reasonably be known by, the licensed physician authorized to declare death. (26 NJSA §6: A1-6, 1991)

FUTURE DIRECTIONS

The standards for the determination of death using brain and circulatory-respiratory tests have been well accepted by medical groups around the world. Approximately 80 countries in the developed and developing world endorse the practice of brain death in medical practice standards or laws or both (Wijdicks, 2002). Nevertheless, there remain local differences in the required specific tests among countries and even among hospitals within countries. Brain death test batteries, for example, have been shown to have a surprising and unjustified degree of variability within American hospitals and therefore remain in need of greater standardization (Bernat, 2008; Greer, Varelas, Haque, & Wijdicks, 2008). Similarly, the tests to determine death using circulatory-respiratory criteria in the context of organ donation show an unjustified variability among hospitals and countries and also are in need of greater standardization (Dhanini, Hornby, Ward, & Shemie, 2012).

Improvements in the standardization of both types of tests are important in their own right, but their development toward increasing precision is being driven strongly by the need for organ transplantation. There is overwhelming acceptance within the organ donation community of the necessity of continuing to respect the "dead-donor rule" to assure sustained public confidence in physicians and organ donation programs. The dead-donor rule, an informal ethical and legal standard that links the Uniform Anatomical Gift Act with state homicide laws, requires that a donor of vital organs (e.g., heart, liver, both lungs, both kidneys) must first be declared dead (Bernat, 2013c; Robertson, 1999).

Formal brain death determinations appear to be declining in incidence in American hospitals because, in many cases, family members and other lawful surrogate decision makers for patients with severe brain injuries and illnesses are ordering the withdrawal of life support measures once the neurological prognosis is determined to be poor—which often occurs before the patient's progression to brain death. As a result, a high percentage of brain death determinations now are performed to permit multiorgan transplantation. The brain illnesses and injuries with poor neurological prognoses that may or may not lead to brain death include traumatic brain injuries, massive intracranial hemorrhages and other large strokes, and neuronal hypoxic-ischemic damage suffered during cardiopulmonary arrest with prolonged impairment in brain oxygenation or circulation.

The need for physicians to develop greater precision in circulatory death determination is linked even more closely to organ donation. Circulatory death determination in patients who are not organ donors usually is performed by simply showing the prolonged cessation of breathing and heartbeat. In the absence of respiratory support and when CPR is foregone or has failed, the permanent cessation of circulation and respiration is sufficient to determine death. However, if the patient is a DCDD donor, to respect the dead-donor rule, the precise moment of death becomes an essential issue. In this setting, greater precision in the death determination is necessary to prove that there will be neither auto-resuscitation nor any medical resuscitative intervention.

The long-standing controversies over the definition and criterion of death will not be resolved fully, but these academic disputes are unlikely to have much effect on physicians' bedside practices in determining death. Whether there will ever be consensus on the precise definition and criterion of death remains a hotly debated topic with cogent arguments on both sides (Bernat, 2013b; Chiong, 2013). To a large extent, the remaining controversies surround what I have termed the paradigm of death: the set of assumptions and conditions that frame the discussion. To some extent, these disagreements may stem from a conceptual dichotomy which the legal scholar and philosopher, Ronald Dworkin, called interpretive versus criterial concepts (Dworkin, 2013). Because it is primarily a biological phenomenon, I regard death as a criterial concept—that is, one for which criteria can be identified. But because the concept of brain death is founded on an idea, namely the irreversible cessation of the organism as a whole, others may plausibly regard it as an interpretive concept that therefore is immune to identifying criteria.

Irrespective of the ongoing academic dispute, most physicians remain unaware of the conceptual controversy and continue to practice according to well-accepted medical and legal guidelines. It is important that medical and legal guidelines for death determination derive from a reasonable biological rationale that is accepted by professionals and the public. Despite the stubborn persistence of the current controversies, I believe the preponderance of evidence shows that the degree of professional and public acceptance for both brain and circulatory death testing satisfies this condition (Bernat, 2006).

NOTES

1. In a little-discussed article, Viennese scholar Raphael Bonelli and colleagues analyzed the specific criteria of life forms and higher organisms (Bonelli, Prat, & Bonelli, 2009). They observed that all life forms have a delimited unity that is characterized by four criteria: (1) dynamics, or signs of life, such as metabolism, regeneration, growth, and propagation; (2) integration, the requirement that the life process derives from the mutual interaction of its component parts; (3) coordination, the requirement that the interaction of the component parts is maintained within a certain order; and (4) immanency, the requirement that the preceding characteristics originate from and are intrinsic to the life form.

   Bonelli et al. then identified four criteria that make a life form a unified whole organism: (1) completion, the requirement that an organism is not a component part of another living entity but is itself an intrinsically independent and completed whole; (2) indivisibility, the condition of intrinsic unity that no organism can be divided into more than one living organism and, if such a division occurs and the organism survives, the completed organism must reside in one of the divided parts; (3) self-reference or auto-finality, the characteristic that the observable life processes and functions of the component parts serve the self-preservation of the whole, even at the expense of the survival of its parts, because the health and survival of the living whole is the primary end in itself; and (4) identity, the circumstance that, despite incremental changes in form and the loss or gain of certain component parts (that even could eventually result in the exchange of all component atoms), the living being remains one and the same throughout life (Bonelli et al., 2009).

Bonelli et al. concluded that the death of an organism is the loss of these four characteristics that render an organism no longer capable of functioning as a whole. They pointed out that in higher animal species, with the irreversible cessation of all functions of the entire brain (brain death), the organism has permanently lost the capacity to function as a whole and therefore is dead. The organism has lost immanency because its life processes no longer spring from itself but result from external intensive care support. The organism has lost auto-finality because whatever control over the component organ subsystem parts that remains now is directed at the level of the surviving parts and no longer at the whole. The organism has lost self-reference because the continued functioning of its parts no longer supports the function of the whole. The organism has lost completeness and indivisibility because its separate component parts and subsystems no longer belong to each other and no longer constitute a whole (Bonelli et al., 2009).

REFERENCES

American Academy of Pediatrics, Committee on Bioethics. (2013). Policy statement: Ethical controversies in organ donation after circulatory death. *Pediatrics*, *131*, 1021–1026.

Beresford, H. R. (1999). Brain death. *Neurologic Clinics*, *17*, 295–306.

Bernat, J. L. (1992). How much of the brain must die in brain death? *Journal of Clinical Ethics*, *3*, 21–26.

Bernat, J. L. (1998). A defense of the whole-brain concept of death. *Hastings Center Report*, *28*(2), 14–23.

Bernat, J. L. (2002). The biophilosophical basis of whole-brain death. *Social Philosophy and Policy*, *19*(2), 324–342.

Bernat, J. L. (2006). The whole-brain concept of death remains optimum public policy. *Journal of Law, Medicine and Ethics*, *34*, 35–43.

Bernat, J. L. (2008). How can we achieve uniformity in brain death determinations? *Neurology*, *70*, 252–253.

Bernat, J. L. (2010a). How the distinction between "irreversible" and "permanent" illuminates circulatory-respiratory death determination. *Journal of Medicine and Philosophy*, *35*(3), 242–255.

Bernat, J. L. (2010b). How auto-resuscitation impacts death determinations in organ donors. *Critical Care Medicine*, *38*, 1377–1378.

Bernat, J. L. (2013a). Controversies in defining and determining death in critical care. *Nature Reviews Neurology*, *9*, 164–173.

Bernat, J. L. (2013b) There can be agreement as to what constitutes human death. In Arthur L. Caplan & Robert Arp (Eds.). *Contemporary debates in bioethics* (pp. 377–387, 397–398). West Sussex, UK: Wiley-Blackwell.

Bernat, J. L. (2013c). Life or death for the dead-donor rule? *New England Journal of Medicine*, *369*, 1289–1291.

Bernat, J. L. (2013d). On noncongruence between the concept and determination of death. *Hastings Center Report, 43*(6), 25-33.

Bernat, J. L., Bleck, T. P., Blosser, S. A., Bratton, S. L., Capron, A. M., Cornell, D., . . . & Shemie, S. D. (2014). Circulatory death determination in uncontrolled organ donors: A panel viewpoint. *Annals of Emergency Medicine, 63*, 384-390.

Bernat, J. L., Capron, A. M., Bleck, T. P., Blosser, S., Bratton, S. L., Childress, J. F., . . . & White, D. B. (2010). The circulatory-respiratory determination of death in organ donation. *Critical Care Medicine, 38*, 963-970.

Bernat, J. L., Culver, C. M., & Gert, B. (1981). On the definition and criterion of death. *Annals of Internal Medicine, 94*, 389–394.

Bernat, J. L., Culver, C. M., & Gert, B. (1982). Defining death in theory and practice. *Hastings Center Report, 12*(1), 5–9.

Bernat, J. L., D'Alessandro, A. M., Port, F. K., Bleck, T. P., Heard, S. O., Medina, J., . . . & Delmonico, F. L. (2006). Report of a national conference on donation after cardiac death. *American Journal of Transplantation, 6*, 281–291.

Bonelli, R. M., Prat, E. H., & Bonelli, J. (2009). Philosophical considerations on brain death and the concept of the organism as a whole. *Psychiatria Danubina, 21*, 3–8.

Capron, A. M., & Kass, L. R. (1972). A statutory definition of the standards for determining human death: An appraisal and a proposal. *University of Pennsylvania Law Review, 121*, 87–118.

Chiong, W. (2013). There cannot be agreement as to what constitutes human death: Against definitions, necessary and sufficient conditions, and determinate boundaries. In Arthur L. Caplan & Robert Arp (Eds.). *Contemporary debates in bioethics* (Chapter 24). West Sussex, UK: Wiley-Blackwell.

Clayton, P., & Kauffman, S. A. (2006). On emergence, agency, and organization. *Biology and Philosophy, 21*, 501–521.

A definition of irreversible coma: Report of the Ad Hoc Committee of the Harvard Medical School to Examine the Definition of Brain Death. (1968). *Journal of the American Medical Association, 205*, 337–340.

Dhanini, S., Hornby, L., Ward, R., & Shemie, S. (2012). Variability in the determination of death after cardiac arrest: A review of guidelines and statements. *Journal of Intensive Care Medicine, 27*, 238–252.

Dworkin, R. (2013). *Justice for hedgehogs.* Cambridge, MA: Belknap Press.

Furton, E. J. (2002). Brain death, the soul, and organic life. *National Catholic Bioethics Quarterly, 2*, 455–470.

Greer, D. M., Varelas, P. N., Haque, S., & Wijdicks, E. F. (2008). Variability of brain death determination guidelines in leading US neurologic institutions. *Neurology, 70*, 284–289.

Halachic Organ Donor Society. (2015). *Halachic issues* [Website]. Retrieved from September 7, 2015. http://www.hods.org/English/h-issues/videos.asp

Hornby, K., Hornby, L., & Shemie, S. (2010). A systematic review of autoresuscitation after cardiac arrest. *Critical Care Medicine, 38*, 1246–1253.

Jain, S., & Maheshawari, M. C. (1995). Brain death: The Indian perspective. In C. Machado (Ed.). *Brain death* (pp. 261–263). Amsterdam: Elsevier.

Koch, C. (2004). *The quest for consciousness: A neurobiological approach* (pp. 10–11). Englewood, CO: Roberts & Company.

Korein, J. (1978). The problem of brain death: Development and history. *Annals of the New York Academy of Sciences, 315*,19–38.
Lock, M. (1995). Contesting the natural in Japan: Moral dilemmas and technologies of dying. *Culture, Medicine and Psychiatry, 19*, 1–38.
Loeb, J. (1916). *The organism as a whole.* New York: G. P. Putnam's Sons.
Mahner, M., & Bunge, M. (1997). *Foundations of biophilosophy* (pp. 29–30). Berlin, Springer-Verlag.
Miles, S. (1999). Death in a technologic and pluralistic culture. In S. J. Youngner, R. M. Arnold, & R. Schapiro (Eds.). *The definition of death: Contemporary controversies* (pp. 311–318). Baltimore: John Hopkins University Press.
Mollaret, P., & Goulon, M. (1959). Le coma dépassé (mémoire préliminaire). *Revue Neurologique, 101*, 3–15.
Munjal, K. G., Wall, S. P., Goldfrank, L. R., Gilbert, A., Kaufman, B. J., & Dubler, N. N. (2013). A rationale in support of uncontrolled donation after circulatory determination of death. *Hastings Center Report, 43*(1), 19–26.
Nakagawa, T. A., Ashwal, S., Mathur, M., Mysore, M. R., Bruce, D., Conway, E. E., Jr., . . . & Child Neurology Study (2011). Guidelines for the determination of brain death in infants and children: An update of the 1987 Task Force recommendations. *Critical Care Medicine, 39*, 2139–2155.
Ogata, J., Imakita, M., Yutani, C., Miyamoto, S., & Kikuchi, H. (1998). Primary brainstem death: A clinic-pathological study. *Journal of Neurology, Neurosurgery, and Psychiatry, 51*, 646–650.
Olick, R. S. (1991). Brain death, religious freedom, and public policy: New Jersey's landmark legislative initiative. *Kennedy Institute of Ethics Journal, 4*, 275–288.
Olick, R. S., Braun, E. A., & Potash, J. (2009). Accommodating religious and moral objections to neurological death. *Journal of Clinical Ethics, 20*, 183–191.
Padela A. I., Arozullah, A., & Moosa, E. (2013). Brain death in Islamic ethico-legal deliberation: Challenges for applied Islamic bioethics. *Bioethics, 27*, 132–139.
Pallis, C. (1983). *ABC of brainstem death.* London: British Medical Journal Publishers.
Parnia, S., Waller, D. G., Yeates, R., & Fenwick, P. (2001). A qualitative and quantitative study of the incidence, features, and etiology of near death experiences in cardiac arrest survivors. *Resuscitation, 48*, 149–156.
Pontifical Academy of Sciences. (2007). *The signs of death.* Scripta Varia 110. Vatican City: Pontifical Academy of Sciences.
President's Commission for the Study of Ethical Problems in Medicine and Biomedical and Behavioral Research. (1981). *Defining death: Medical, legal and ethical issues in the determination of death.* Washington, DC: U.S. Government Printing Office.
Robertson, J. A. (1999). The dead donor rule. *Hastings Center Report, 29*(6), 6–14.
Rosner, F. (1999). The definition of death in Jewish law. In S. J. Youngner, R. M. Arnold, & R. Schapiro (Eds.). *The definition of death: Contemporary controversies* (pp. 210–221). Baltimore: Johns Hopkins University Press.
Shewmon, D. A. (2009). Brain death: Can it be resuscitated? *Hastings Center Report, 39*(2), 18–24.
Shewmon, D. A. (2010). Constructing the death elephant: A synthetic paradigm shift for the definition, criteria, and tests for death. *Journal of Medicine and Philosophy, 35*, 256–298.

Task Force on Death and Dying of the Institute of Society, Ethics, and the Life Sciences. (1972). Refinements in the criteria for the determination of death: An appraisal. *Journal of the American Medical Association, 221*, 48–53.

Veatch, R. M. (1993). The impending collapse of the whole-brain definition of death. *Hastings Center Report, 23*(4), 18–24.

Veatch, R. M. (2005). The death of whole-brain death: The plague of the disaggregators, somaticists and mentalists. *Journal of Medicine and Philosophy, 30*, 353–378.

Wijdicks, E. F. M. (2002). Brain death worldwide: Accepted fact but no global consensus in diagnostic criteria. *Neurology, 58*, 20–25.

Wijdicks, E. F. M., Varelas, P. N., Gronseth, G. S., & Greer, D. M. (2010). Evidence-based guideline update: Determining brain death in adults. *Neurology, 74*, 1911–1918.

5

Modes of Consciousness

TIM BAYNE AND JAKOB HOHWY

Philosophical interest in consciousness has tended to focus on specific states of consciousness—for example, the kinds of conscious states that one is in when experiencing pain, looking at a sunset, or feeling depressed. Theorists have discussed whether the phenomenal character of such states can be captured by appeal to their intentional contents; they have considered why such states might be conscious at all; and they have explored ways in which such states might be accommodated within a physicalist conception of the world. This focus has led to much progress in understanding the nature of specific conscious states, but this progress has been accompanied by the neglect of many other components of consciousness. In particular, theorists have paid scant attention to what we will here call *modes of consciousness* (and elsewhere call *global states of consciousness*).

It is not possible to provide a straightforward definition of a mode of consciousness, but to a first approximation we might describe modes of consciousness as global ways of being conscious. As with many aspects of consciousness, modes are perhaps best approached by means of examples. The mode of consciousness with which we are most familiar is the waking state. Other modes of consciousness are associated with:

- Rapid eye movement (REM) dreaming (Hobson, 1999)
- The hypnotic state (Jamieson, 2007)
- The state of (light) anesthesia (Alkire & Miller, 2005)
- The minimally conscious state (MCS) (Giacino, Kalmar, & Whyte, 2004; Giacino et al., 2009; Laureys, 2005)
- Stupor (Posner, Saper, Schiff, & Plum, 2007)

- Epileptic absence seizures (Englot & Blumenfeld, 2009)
- Dementia (Posner et al., 2007)
- Delirium (Bhat & Rockwood, 2007)

In fact, there are entire frameworks for taxonomizing disorders of consciousness that appeal to distinctions between what we call modes of consciousness (Demertzi et al., 2008).

The challenges posed by accounting for the modal nature of consciousness are quite different from those posed by specific conscious states. In the latter case, the object of enquiry is reasonably clear. We have at least a rough idea of what distinguishes specific conscious states as a class and how to individuate one kind of specific conscious state from another. The chief puzzles posed by specific conscious states concern how their phenomenal character can be explained in terms of their neurofunctional and representational properties. By contrast, the challenges posed by conscious modes concern their very identification. It is not obvious what a mode of consciousness *as such* is, nor is it obvious what distinguishes one mode of consciousness from another. Indeed, it is unclear whether all of the phenomena listed above do indeed involve distinct modes of consciousness.

It is useful to draw a distinction between two kinds of approaches that one can adopt with respect to the modal structure of consciousness—what we call a *taxonomic* approach and an *explanatory* approach. As the label suggests, a taxonomic approach attempts to chart the conceptual contours of modes. It aims to identify the ways in which modes as a class differ from other states of consciousness and how one mode of consciousness differs from another. An explanatory approach to modes goes beyond taxonomy and attempts to identify the neurofunctional mechanisms that are responsible for the modal nature of consciousness. (Of course, the line between these two approaches is not hard and fast. Any attempt to provide an explanatory model of modes presupposes a certain taxonomy, and an explanatory model may in turn have implications for taxonomy. But as a rough approximation, the distinction is a useful one.)

With respect to the explanatory approach, there is a significant body of work in the neurosciences examining the neurofunctional processes underpinning various modes of consciousness, in particular sleep and related conditions. Putting this literature to one side, we will focus here on taxonomic questions. Our primary interest is with the more abstract question of how modes of consciousness as such should be understood. Developing such an account is the primary aim of this chapter.

We assume that a taxonomic account of modes ought to meet the following three desiderata. First, it must preserve the distinction between modes of consciousness and specific conscious states. The contrast between, for example, the mode of consciousness associated with normal wakefulness and that associated with REM dreaming is very different from the contrast between the state of consciousness associated with pain and that associated with hearing a melody. Similarly, disorders of consciousness that involve alterations in a creature's mode of consciousness (e.g., epileptic absence seizures, the MCS) are quite unlike those that involve alterations in a creature's content-specific states (e.g., blindsight, unilateral neglect, anosognosia). Although an account of modes might well have implications for accounts of specific conscious states (and vice-versa), it is highly implausible to suppose that conscious modes can be identified with specific conscious states (or complexes thereof).

A second constraint on an account of modes is that it must capture the fact that modes are *global* states.[1] Modes characterize the overall conscious state of a creature. However, the fact that modes are global states does not mean that they exclude each other. *Some* modes certainly exclude each other. For example, being in the vegetative state excludes being in the minimally conscious state and vice-versa. Similarly, being anaesthetized excludes being in the state of normal wakefulness and vice-versa. But some modes do not exclude each other. For example, we see no reason why a minimally conscious state patient couldn't undergo an absence seizure.[2]

A third constraint on an account of modes is that it must do justice to the fact that modes admit of some kind of ordering. Intuitively, the modes of consciousness that occur in the MCS, epileptic absence seizures, delirium, and REM dreaming (assuming that these conditions do involve consciousness) are in some intuitive sense "lower" than the mode of consciousness that characterizes ordinary waking consciousness. However, there are many open questions about the dimensional structure of conscious modes, and we certainly should not assume that modes can be structured in terms of a complete ordering. For example, it might turn out that there are modes A, B, and C such that A and B are both higher than C, yet there may be no determinate ranking of A and B relative to each other.

In this chapter, we first provide a brief overview of how modes are treated in the current literature, and then consider reasons for thinking that there is no full adequate account of modes on the market. In the following sections, we explore three approaches to modes: a content-based approach, a structural approach, and a functional-role approach. Although none of these approaches is able to provide an adequate account of modes on its own, when taken together they provide us with the core of an analysis of modes. Finally, we turn to the question of whether there are ways to unify the three approaches.

MODES, LEVELS, AND BACKGROUND STATES

Although the modal nature of consciousness has not received a great deal of attention from either scientists or philosophers, it has not been entirely overlooked. Discussion of the notion—or at least something very much like it—can be found in connection with what philosophers have called "background states of consciousness" and in connection with what clinicians have called "levels" of consciousness. We begin with the former notion.

Here is how David Chalmers (1998) introduced the notion of a background state:

> A background state is an overall state of consciousness such as being awake, being asleep, dreaming, being under hypnosis, and so on. Exactly what counts as a background state is not entirely clear, as one can divide things up in a number of ways, and with coarser or finer grains, but presumably the class will include a range of normal and of "altered" states. . . . Creature consciousness is the most coarse-grained background state of consciousness: it is just the state of being conscious. Background states will usually be more fine-grained than this, but they still will not be defined in terms of specific contents or [sensory] modalities. (p. 18)

The examples that Chalmers gives of background states correspond very closely to what we regard as paradigm cases of modes, and it is reasonable to think that his background states of consciousness are equivalent to our modes of consciousness. However, there is one claim in this passage that we would want to distance ourselves from. It is not clear to us that "creature consciousness" qualifies as a mode of consciousness, even with the proviso that it is the most coarse-grained of conscious modes. Creature consciousness, as Chalmers uses the phrase, is simply the property of being conscious.[3] Modes of consciousness, by contrast, are determinates of creature consciousness—they are *ways* of being conscious. One cannot be conscious *simpliciter*, but must be conscious by being in a state of (e.g.) normal wakefulness, REM dreaming, or epileptic absence seizure. It may be useful to think of consciousness in terms of determinable-determinate hierarchies (to illustrate, *red* is a determinate of the determinable *color*, and *scarlet* is in turn a determinate of *red*; being scarlet is thus a way of being red and of being colored). Intuitively, it seems correct to say that being in a mode of consciousness is a way of being conscious; similarly, it seems correct to say that having a content presented in consciousness is a way of being conscious. (Note, however, that even though both modes and content lend themselves to being described in terms of determinable-determinate hierarchies, it is not obvious how the two hierarchies are connected, for contents and modes do not bear determinate-determinable relations to each other.)

A further reservation we have with Chalmers's discussion is that the terminology of "background states" is potentially misleading. The phrase might suggest that such states are phenomenally recessive in the way that bodily sensations (e.g., queasiness) and affective experiences (e.g., moods) often are—that they occupy the unattended periphery of the stream of consciousness. But although certain modes might be associated with distinctive background phenomenology, they are not themselves components of the stream of consciousness in the way that sensations or affective experiences are, and the contrast between content and mode should not be confused with the contrast between the phenomenal focus and phenomenal background.

A more influential approach to modes—especially within neurology—appeals to the notion of a *level* of consciousness. Although the term itself does not appear in the following passage from Zeman (2001), the idea that consciousness takes different levels is clearly at work in it:

> In everyday neurological practice consciousness is generally equated with the waking state, and the abilities to perceive, interact and communicate with the environment and with others in the integrated manner which wakefulness normally implies. Consciousness in this sense is a matter of degree: a range of conscious states extends from waking through sleep into coma. These states can be defined objectively, using behavioural criteria like those supplied by the Glasgow Coma Scale . . . Thus we speak of consciousness dwindling, waning, lapsing and recovering; it may be lost, depressed, regained. (p. 1265)

Although some neurological discussions suggest that modes of consciousness can be identified with distinct levels of consciousness, the standard view appears to be that modes are regions within a two-dimensional space. One axis of this space is labeled "contents of consciousness" or "awareness"; the other axis is labeled "levels of consciousness," "wakefulness," or "arousal." This conception of modes is implicit in figure 5.1, versions of which can be found in many recent discussions of consciousness (e.g., Demertzi et al., 2008; Laureys, 2005; Mormann & Koch, 2007).

This diagram raises a number of important questions. Perhaps the most important question concerns how exactly we are to understand the two axes. Consider first the vertical axis, which is here labeled "Content of consciousness (awareness)." What distinguishes creatures that are located at the top of this axis from those that are located near the middle or the bottom of the axis? Perhaps the idea is that creatures that are in conscious modes located toward the top of the vertical axis enjoy a wider range of conscious contents than do creatures that are in modes located toward the bottom of this axis. In the latter case, consciousness would be either entirely absent or limited to a narrow range of primitive contents (e.g., low-level perceptual experiences or bodily sensations), whereas in the former case, consciousness would be characterized by a wide range of complex contents. We will return to this idea in due course.

The horizontal axis of this diagram is rather more problematic. For one thing, it is unclear what precisely the axis is meant to represent. Labeling this axis "Level of consciousness" suggests that there is some single property—"consciousness"—that admits of degrees, and that modes that occur toward the right end of this axis exemplify more of this property than do modes that occur toward the left end of the axis. We regard this proposal as problematic, for it is far from clear that consciousness does come in degrees. Arguably, being conscious is akin to the property of being a parent: A creature either instantiates the property or it does not. (Someone who has seven children is not more of a parent than someone who has only one child.) And even if consciousness does come in degrees, it is very doubtful whether the distinctions that are being tracked by the horizontal axis correspond to differences in degrees of consciousness, for the syndromes that appear in the bottom right-hand corner of the matrix (sleepwalking, complex partial and absence seizures, and the vegetative state) are clearly not conditions in which subjects have high degrees of consciousness—Indeed, these are syndromes in which consciousness is typically assumed to be completely absent! Moreover, even if consciousness is sometimes present in these conditions, it is implausible to suppose that the degree of consciousness that occurs in them is comparable to that which occurs in normal waking awareness. In short, the horizontal axis of Figure 5.1 is not usefully thought of in terms of levels of consciousness.

Might it be more illuminating to describe this dimension in terms of the label that appears in brackets—"wakefulness"? Roughly speaking, we can think of wakefulness in terms of orientation to one's environment. This proposal looks promising insofar as categorizing modes of consciousness in terms of their degree of wakefulness captures the fact that modes that involve high degrees of environmental orientation (such as ordinary conscious wakefulness) are intuitively "higher" than modes that involve only partial or selective environmental orientation, such as hypnosis, REM dreaming, or the MCS (although it is not immediately obvious how one should order these latter three modes relative to each other).

But there are also problems involved in conceiving of the horizontal axis in terms of wakefulness. Perhaps most noticeably, this interpretation is at odds with the location of the syndromes at the bottom right of Figure 5.1. Sleep-walking and complex partial and absence seizures involve only impoverished degrees

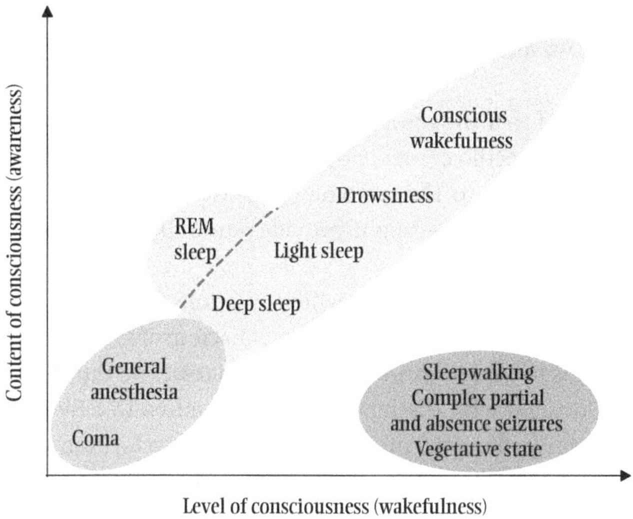

Figure 5.1 A common way of illustrating in simplified form the main components of consciousness. This figure was adapted from Laureys, 2005, and Mormann and Koch, 2007. (Mormann and Koch placed the vegetative state and the minimally conscious state in the bottom left corner and REM sleep more toward the upper left corner; they also labeled the horizontal axis "Arousal").

of orientation to one's environment, and the vegetative state does not seem to involve *any* significant environmental orientation at all. So if these syndromes are correctly represented as having high degrees of the property represented by the horizontal axis then we cannot think of that property in terms of orientation to one's environment. Similar considerations apply to the other notion sometimes used to label the horizontal axis, namely "arousal."

The hope represented by Figure 5.1 is a laudable one, for it would certainly be an advance in our understanding of modes if they could be fully captured by appeal to a two-dimensional matrix. However, it is far from clear that modes will succumb to this kind of reductive analysis. At the very least, we have suggested that there are a number of open questions about the treatment of modes suggested by Figure 5.1. We will revisit the notions of wakefulness and arousal later in this chapter in connection with the question of whether appeals to these notions can unify the various features of modes.

CONTENT-BASED APPROACHES TO MODES

We turn now to the task of developing an abstract account of modes of consciousness. In what follows, we will examine content-based, structure-based, and functional approaches to modes, arguing that each of these approaches is limited but also that each needs to be incorporated as part of an overall view. In this section, we focus on approaches to modes that appeal to the *contents* of consciousness.

We have argued that the modal structure of consciousness must be distinguished from the specific contents of consciousness, but this leaves open the possibility that appeals to the contents of consciousness—and in particular *relations* between contents—might provide some illumination of the modal structure of consciousness.

There are a number of ways in which relations between contents might be implicated in the modal structure of consciousness. One content-based approach appeals to the *bandwidth* of consciousness, and the idea that distinct modes are associated with distinct conscious bandwidths. By the bandwidth of consciousness we mean the number of distinct objects, events, or properties that can be represented in consciousness at any one point in time (or across short intervals). Estimates of the bandwidth of the normal waking state vary widely, but it is often suggested that consciousness can normally accommodate only four items (e.g., Cowan, 2000).

An approach to modes that appeals to the bandwidth of consciousness meets all three of the desiderata that we outlined in the introduction. It provides a clear distinction between the modes of consciousness and the contents of consciousness, for although bandwidth is defined over contents it is independent of any particular type of content. Appealing to bandwidth also accounts for the fact that modes are global features of consciousness, for the bandwidth of consciousness is a property of a creature's overall stream of consciousness. And the bandwidth approach also promises to capture the dimensional nature of modes of consciousness insofar as some modes might involve a wider bandwidth in consciousness than others.

But despite these points, it is unclear just how much light an appeal to the bandwidth of consciousness sheds on the modal nature of consciousness. For all we know, it is entirely possible that creatures occupying very different modes of consciousness could have the same conscious bandwidth. It also seems possible that two individuals might occupy the same mode of consciousness but differ in the bandwidth of their stream of consciousness. Our current ignorance concerning the bandwidth of consciousness makes it very difficult to say how likely it is that either of these possibilities are realized, but the mere fact that they are both intelligible indicates that the notion of a mode of consciousness cannot be fully captured by appeal to claims about the bandwidth of consciousness. Moreover, even though there is some intuitive plausibility in the idea that "lower" modes of consciousness have a smaller conscious bandwidth than "higher" modes of consciousness, to the best of our knowledge there is very little evidence in support of this proposal, and it is not clear that any intuitive ordering between modes maps onto differences in bandwidth.[4]

A more promising content-based approach to modes of consciousness appeals to the idea that certain modes involve restrictions in the kinds of contents that can be represented in consciousness. We will call this the *limiting* approach to modes. Patients in a MCS may be able to consciously experience certain types of contents—for example, elementary sensory features such as loudness, hue, brightness, thirst, or pain—but it seems unlikely that they are capable of consciously representing complex features of their environment, such as the identity of particular individuals or the implications of a certain social cue. This notion of a limited range of conscious content is supported by evidence suggesting that the mechanisms responsible for "raw emotional feelings" are often preserved in disorders of consciousness even though the capacities required for more cognitively complex states are disrupted (e.g., Panksepp, 2005; Panksepp, Fuchs, Garcia, & Lesiak, 2007). It is also possible that some psychotic states, such as those induced by hallucinogenic drugs, involve the lifting of ordinary limits on conscious contents, with the result that subjects enjoy an *expansion* in the range of perceptual and cognitive states that are accessible to consciousness (Geyer & Vollenweider, 2008), thus justifying (to some extent at least) the intuitive thought that such modes of consciousness are "higher" than ordinary waking awareness.

Cognitive neuroscience indicates that neural networks specialize in particular domains of content (Greicius, Krasnow, Reiss, & Menon, 2003). Modes appear to limit consciousness by taking some of these networks "off-line." For example, there is evidence that activity in the default mode network—a network that appears to specialize in representation of the subject's "internal" environment and self-related tasks such as envisioning one's future (Buckner, Andrews-Hanna, & Schacter, 2008; Gusnard, 2005)—is abolished in coma but gradually increases as subjects migrate from the vegetative state to the MCS and then to the state of full waking awareness (Boly et al., 2008). In some cases (e.g., REM sleep, anesthesia), networks may be taken off-line only transiently, whereas in other cases (e.g., the MCS), they are permanently disabled.

Although reference to content-limiting appears to capture certain aspects of the modal nature of consciousness, it is unlikely that it will provide us with a full analysis of modes. One cannot draw inferences about what conscious mode a creature is in by merely looking at the limitations that govern its conscious contents. In fact, there are two kinds of limits on the contents of consciousness that are independent of modes. First, a creature may lack the perceptual and cognitive capacities required for representing certain types of contents in consciousness. The fact that human infants and nonhuman animals are able to consciously represent only some of the contents that are available to neurotypical adult human beings is obviously not to be explained by appeal to the modal structure of consciousness.[5]

Limits on the contents of consciousness can also arise as a result of content-specific deficits in consciousness. Consider a subject who has suffered from a number of lesions, each of which limits content in a particular domain. We can imagine, for example, an individual with akinetopsia, unilateral neglect, word deafness, and chronic deafferentation. Such a person would undergo a severe restriction in the range of contents that can be experienced but would not thereby be in a distinctive mode of consciousness. On the contrary, there is every reason to think that such a person could be in the same range of conscious modes that are available to neurotypical individuals. Indeed, it seems possible—albeit, perhaps, highly unlikely—that the conscious contents of an individual in a state of ordinary wakefulness could be constrained in exactly the same ways that the contents of an MCS patient are constrained. The upshot of these considerations is that although information about a person's conscious mode might support inferences about the ways in which its conscious contents are constrained, we cannot so easily draw inferences about conscious mode from information about the ways in which its conscious contents are constrained.

STRUCTURAL APPROACHES TO MODES

A different approach to modes focuses not on the contents of consciousness but on the phenomenal *structure* of consciousness. As with content-based approaches to consciousness, there are various aspects of the structure of consciousness that might be implicated in an account of modes.

One such aspect is temporal structure. Consider the specious present, a classic illustration of which is provided by musical perception. Although the auditory system is exposed to only a single note at a time, our experience of music is broader than this and may be likened to a moving window in which the representation of previously presented notes (retention) and the anticipation of future notes (protention) is included in a single experiential "now." There is some reason to suspect that the duration of the specious present varies depending on a creature's mode of consciousness (Pockett, 2003). If the psychophysical construct of "duration perception" is taken as an indirect guide to the structure of the specious present, differences can be found in schizophrenia, Parkinson's disease, autism, and attention deficit hyperactivity disorder (ADHD) (see Allman & Meck, 2012, and Grondin, 2010, for reviews). There is also reason to think that the specious present is disrupted in delirium, which is characterized by arousal-related clouding of consciousness, attention deficits, and difficulties in the perception of temporal order (Bhat & Rockwood, 2007).

Appealing to the temporal structure of consciousness meets the three desiderata that we previously identified: It provides a clear contrast between modes and contents; it promises to capture the fact that modes are global features of consciousness insofar as the duration of the specious present might be relatively invariant across different types of contents; and it may explain some of the dimensional structure of consciousness insofar as some modes might involve a longer specious present than others do. But these points notwithstanding, it should also be recognized that any attempt to account for modes by appealing to temporal structure will be highly speculative at present. We know rather little about the structure of the specious present in the context of normal wakefulness, and we know even less about its structure in abnormal modes of consciousness. Although it seems likely that "lower" modes of consciousness will involve abnormalities in the temporal structure of consciousness, our judgments about the ordering relations between modes is unlikely to be hostage to discoveries about such matters.

Rather than looking to the temporal structure of consciousness for an account of modes, it might be more promising to consider its *attentional* structure—that is, the ways in which the field of consciousness is distributed between what falls within focal attention and what falls outside it. Ordinary waking experience involves a distinction between contents (e.g., objects, properties) that fall within the focus of attention and those that "surround" it (Kriegel, 2004; Watzl, 2012).[6] The selection of contents for attentional focus involves a delicate interplay between endogenous and exogenous factors. A loud sound will typically capture one's attention, but one can also choose to ignore attention-grabbing features of one's environment and instead attend to stimuli that, despite not being exogenously salient, are relevant to one's current task.

There are two ways in which the attentional structure of consciousness might bear on the analysis of modes. One possibility is that modes might be distinguished from each other in terms of how attention structures the field of consciousness. For example, one can imagine a conscious field in which there is no gradient of attention, either because everything falls within focal attention or because nothing falls within focal attention. Attentional is thought to be disrupted in many global disorders of consciousness (Schiff & Plum, 2000). A possible example of a restrictive attentional structure concerns the detection of deviations from global (as opposed to local) auditory patterns. Evidence from healthy individuals indicates that the detection of global deviants requires attention. In general VS patients seem not to detect such deviants (although see King et al., 2013), which indicates that the disruption of attention is disrupted in such patients (Bekinschtein et al., 2009). (By contrast, MCS patients are much more likely to show evidence of detecting violations of global regularities.) A possible example of an overinclusive attentional structure is heightened attentional salience induced by drugs such as ketamine (Corlett, Honey, & Fletcher, 2007).

Another possibility is that modes might be distinguished from each other not because of the way in which attention is distributed across the field of consciousness but because of how it is *controlled*. For example, we can imagine a subject in which the distribution of attention is entirely under exogenous control. This description may apply to certain patients in a vegetative state (or at least a behaviorally vegetative state) and patients in MCS, who appear able to form intentions in response to commands and other forms of environmental input without being able to form them endogenously (Owen, 2008). A different impairment of attention occurs in the context of epileptic seizures, in which subjects often experience difficulty controlling both the scope (i.e., broad versus narrow) and the focus (i.e., inner states versus environmental states) of attention (Johanson, Revonsuo, Chaplin, & Wedlund, 2003). It might be appropriate to describe these subjects as exemplifying nonstandard modes of consciousness insofar as the control of attention has been disrupted.

How might the attention-based approach to modes meet the three desiderata that we earlier identified? Thinking of modes in terms of the structure of attention gives us a clear contrast between modes and contents, for although the deployment of attention has implications for content it is not itself a type of conscious content.[7] The attention-based approach also does justice to the fact that modes are global features of consciousness, for both the distribution and control of attention are properties that characterize a creature's overall state of consciousness. The attention-based approach may also be able to provide at least a partial vindication of our intuitions concerning the ordering relations between modes. The state of normal wakefulness qualifies as a "high" mode of consciousness insofar as the control of attention involves a functionally useful distribution of attention across the subject's stream of consciousness and there is an appropriate balance between exogenous and endogenous control, whereas "abnormal" modes of consciousness will qualify as "lower" insofar as the distribution of attention or control of attention (or both) is functionally compromised.

Having said all this, it is doubtful whether appeals to attention can provide a full account of the modal structure of consciousness. Although there are reasons to suspect that the distribution and control of attention is likely to be disordered in abnormal modes of consciousness, we know very little about the nature of attention in such conditions, and the judgment that such modes are "lower" than the mode of everyday waking awareness should not be hostage to the assumption that these conditions involve disturbances in the allocation and control of attention. Attention is likely to play an important role in the analysis of modes, but it will need to be supplemented by an appeal to other factors.

FUNCTIONAL APPROACHES TO MODES

In the previous two sections, we examined ways in which one might appeal to the contents and structure of consciousness in order to illuminate its modal nature. We turn now to an approach that focuses on functional considerations.

At the core of a functional approach to modes is the idea of a consuming system. As the name suggests, a consuming system is a system that employs specific conscious states in the service of some cognitive or behavioral task. There are consuming systems implicated in voluntary motor behavior, verbal report, memory consolidation, decision making, and other tasks. We will leave to one side the important but delicate question of how best to individuate consuming systems; presumably, such systems can be individuated at various levels of grain.

How might the notion of a consuming system capture the modal structure of consciousness? The rough idea is that reference to modes of consciousness provides a convenient way of capturing facts about how the contents of consciousness can drive the creature's consuming systems. Normal wakefulness is characterized by the fact that the contents of consciousness are available to a wide range of consuming systems, whereas in nonstandard modes of consciousness, the contents of consciousness are not widely available for cognitive and behavioral control. For example, creatures in a state of REM dreaming or extreme tiredness are not typically able to use the contents of their conscious states to guide a wide range of behaviors because these states involve disruption to the systems responsible for decision making, working memory, motor control, and memory consolidation (Durmer & Dinges, 2005; Windt, 2011).

It is important not to confuse the functional approach to modes with functionalist accounts of consciousness as such. Functionalist accounts of consciousness hold that what it is for a mental state to be conscious is for it to play a certain functional role within the cognitive economy of its subject. (An influential example of the functionalist approach is provided by global workspace theories of consciousness, according to which a mental state is conscious in virtue of the fact that it occurs within a central cognitive workspace.) It is possible to endorse a functional account of conscious modes while rejecting functionalist accounts of consciousness, and vice-versa. Nevertheless, there is a natural affinity between functional accounts of consciousness and functional accounts of conscious modes, and the attractiveness of the former lends a certain plausibility to the latter.

How does the functionalist approach fare when measured against our three desiderata? It is clearly able to capture the intuitive contrast between modes of consciousness and specific conscious states, for two creatures could be in precisely the same specific conscious state while differing in conscious mode simply because the contents of these states are available to a different range of consuming systems. For example, the conscious contents of one creature might be available for verbal report, voluntary behavioral control, and memory consolidation, whereas the same conscious contents might be available only for memory consolidation in another creature.

The functional approach also captures the fact that modes are global features of consciousness. Although it is conceptually possible that failures of accessibility might occur in a content-specific manner, as a matter of fact they appear to occur en masse. In other words, if the contents of one sensory modality are unavailable for verbal report, it is very likely that the contents of other sensory modalities will also be unavailable for verbal report. Failures of accessibility typically occur because the consuming system itself is off-line, rather than because the information channel between a specific conscious state and a consuming system has been disrupted.

What about the dimensional aspect of modes? Is the functional approach able to capture order relations between modes? It is; in fact, it is arguable that the functional approach does better in this regard than the other approaches we have considered. As already noted, one of the features of the state of normal waking awareness is that the contents of consciousness are able to drive a wide range of behaviors. This mode of consciousness is associated with the flexible and intelligent pursuit of goals. By contrast, "nonstandard" modes of consciousness are associated with relatively "rigid" and "dumb" forms of behavior, in which subjects are unable to deploy the full suite of consuming systems that they possess. For example, some MCS patients might be able to use their perceptual experiences to pick up an object, but not be able to use them to engage in communication. Similarly, individuals in the grip of an epileptic seizure might be able to use their perceptual experiences to navigate around an obstacle but be unable to use those same experiences to make novel inferences.

However, the functional approach may not be able to provide us with a *complete* ordering of modes. In a number of cases, we might be able to say that mode A differs from mode B in terms of which consuming systems are on-line, but there may be no sense in which A is higher or lower than B. To appreciate this point, imagine a creature with six consuming systems, numbered 1 through 6. In mode A, systems 1 through 5 might be on-line, whereas in mode B, systems 2 through 6 might be on-line. In such a case, there might be no way of ordering modes A and B relative to each other, although both modes might be "higher" than modes in which, for example, only system 4 is on-line.[8]

So, the functional approach looks promising. However, it also faces a number of challenges. One challenge derives from the need to distinguish failures of cognitive and behavioral access that derive from changes in the mode of consciousness from failures caused by other factors. There is an intuitive difference between a creature in which a consuming system has been disrupted by localized brain damage and a creature in which that same consuming system has been disrupted because of a change in conscious mode. For example, we can compare a patient who is unable to report his experiences because he is in an epileptic seizure with a patient who cannot report his experiences because he has aphasia. In the former case, the functional impairment is caused by a change in conscious mode, whereas in the latter case it results from a lesion to the language system. This example suggests that although information about a creature's conscious mode might support inferences about which consuming systems are (or are not) on-line, information about a creature's consuming systems may not support inferences about its conscious mode.

A second challenge facing the functionalist approach to modes is that it threatens to cut modes too finely, or at least in ways that do not correspond to our intuitive conception of the boundaries between modes. Consider, for example, epileptic absence seizures. Arguably, this condition corresponds to a single conscious mode, one that is likely to differ from the mode of consciousness seen in other forms of epileptic seizure and indeed in other pathologies of consciousness (e.g., delirium, MCS). However, there is a significant amount of individual variation in the ways in which absence seizures disrupt a patient's cognitive and behavioral capacities (Blumenfeld, 2005). So, a potential implication of the functionalist approach to modes is that modes should be individuated more finely than we tend to assume pre-theoretically. Whether or not this is an objection to the functionalist approach depends on how much importance ought to be assigned to the pre-theoretical judgment that absence seizures involve a single conscious mode. Arguably, judgments concerning the boundaries between modes should form only the starting point of our enquiry, and we should allow them to be revised in light of theoretical considerations. Moreover, the functional approach to modes may be able to capture the thought that absence seizures involve a single mode of consciousness even if there is significant individual variation in the cognitive and behavioral effects of absence seizures. Rather than identify modes with *points* in functional space (so to speak), we could think of them in terms of *regions* of functional space. The functionalist approach to modes might be able to preserve the intuition that absence seizures involve a single conscious mode as long as the functional consequences of all such seizures fall within this region.

Overall, the functional approach seems somewhat more plausible than the content-based and structural approaches to modes. However, it cannot easily explain the intuitive distinction between a difference in access that stems from a difference in modes and a difference in access that stems from factors unrelated to modes of consciousness. One way to accommodate this distinction would be to appeal to the other two approaches. Thus, it remains a possibility that two or three approaches contribute jointly to the overall manifestation of some modes, or perhaps some common cause modulate all three aspects of modes. Further, it remains a possibility that there are causal interactions between content-related, structural, and functional aspects of modes. For example, it may be that limits on content, or disruption to attentional structure, influence the workings of some cognitive consumer systems, which may be optimized to a particular range and structure of content. Likewise, it may be that disruption to consuming systems feeds back and causes disruptions to the range and the temporal and attentional structure of an agent's contents.

A UNIFYING APPROACH TO MODES?

The three approaches to modes that we have explored are conceptually quite distinct, and it is possible that a creature could be in a "high" mode of consciousness according to one approach but a "low" mode relative to the other approaches. But although such dissociations might be *conceptually* possible, it is a further question whether the full array of modes that is conceptually possible can actually occur in practice. The answer would seem to be no. In fact, large areas of "modal space" would seem to be completely uninhabited. For example, one does not find creatures who are able to enjoy an unbounded, normally structured range of conscious content but have access to only a few primitive consuming systems; nor does one find creatures who have access to the wide range of consuming systems that characterize normal waking consciousness but can enjoy only a very restricted range of primitively structured sensory states. Indeed, in general, there seems to be a rough correlation between the "contentful," "structural," and "functional" aspects of a creature's overall conscious state. (However, there are exceptions to this correlation, as we shall discuss below.) How might this feature of the modal structure of consciousness be accounted for?

Perhaps the most straightforward explanation would involve an appeal to the operation of a single system. High levels of activity in this system would be a common cause of fewer limits on the types of contents that can enter consciousness, of optimal structure within the stream of consciousness, and of widespread availability of conscious content to consuming systems. In contrast, low levels of activity in this system would result in restrictions along all three dimensions. This would be a significant result because, a priori, there is no reason to assume that there is a single system responsible for the modal structure of consciousness. After all, modes could turn out to be nominal rather than real kinds—artificial constructs that emerge from the interaction of a number of fundamentally distinct systems.

What might the system that supplies this common causal ground be? Textbook presentations of consciousness within clinical neuroscience often suggest that a creature's overall level of consciousness is grounded in the state of its *arousal* system (e.g., Frackowiak et al., 2004; Laureys, 2005; Posner et al., 2007; Zeman, 2001). Indeed, underlying the view that "consciousness has two major components: content and arousal" (Posner et al., 2007, p. 5; see also Figure 5.1), one often finds the idea that arousal is more basic. This idea is captured in the following passage (Frackowiak et al., 2004):

> The contents of consciousness can vary quite independently of the level of consciousness [i.e., arousal]. Specific brain lesions can alter the contents of consciousness without having any effects on the level of consciousness. . . . On the other hand the level of arousal has a major influence on the contents of consciousness. On the whole, as arousal increases the extent and quality of conscious experience also increases. (p. 270)

According to this view, content, structure, and access have no causal influence on arousal, whereas arousal has a causal influence on content, structure, and access. In other words, arousal is suggested as a mechanistic common cause which controls the features of modes we have identified.

But what precisely *is* arousal? Arousal is often explicated by appeal to behavioral capacities. For example, "Arousal refers to the level of alertness . . . clinically it is indicated by opening of the eyes" (Posner et al., 2007, p. 5). Such behavioral explications of arousal are often very rough and ready, and have a somewhat pragmatic flavor: "Arousal is defined as the degree of sensory stimulation required to keep the patient attending to the examiner's questions" (Posner et al., 2007, p. 183). This pragmatism is reflected in the clinical tools designed to assess levels of consciousness (e.g., Glasgow Coma Scale, Coma Recovery Scale-Revised), although the limitations of this approach are often acknowledged: "It is difficult to equate the lack of motor responses to the depth of the coma, as the neural structures that regulate motor responses differ from those that regulate consciousness, and they may be differentially impaired by specific brain disorders" (Posner et al., 2007, p. 7). If arousal is understood in these behavioral terms, then it seems at least partly affiliated with the functional aspects of the modes we have indicated. Presumably, the extent to which one can orient oneself in a properly responsive manner toward the environment depends on the degree to which systems of cognitive and behavioral control are on-line. And if that is right, then arousal would not really be independent of the aspects of modes it was meant to explain.

An alternative conception of arousal involves an appeal to neurological properties, such as features of the brainstem ascending arousal system (capturing, for example, the way in which the ventrolateral preoptic system can inhibit the ascending arousal system during sleep; see Posner et al., 2007, p. 23). The idea here is that different types of activity in these systems might explain patterns of change in the aspects of modes we described earlier. This would provide a neural candidate for the common cause of fluctuations in these features of modes of consciousness. Of course, it would still be an open empirical question whether activity in this system might explain the various modal aspects of consciousness that we have identified. One reason to think that arousal might not be able to play this unifying role is that there are conditions in which an individual's mode of consciousness appears to dissociate from its state of arousal. In fact, there are two kinds of problem cases for the arousal-based analysis. One kind stems from conditions (e.g., MCS) in which individuals can be in a state of high arousal despite being in what one would intuitively characterize as a low mode of consciousness. The vegetative state is even more problematic here, for patients in this state can be highly aroused but consciousness is usually thought to be altogether absent, and even where it may be present it is arguably present only at a very low level (i.e., with severe limits on content, structure, and access). Further challenges in this vein are posed by sleep walking (Mahowald & Schenck, 1992), complex partial seizures (Englot & Blumenfeld, 2009), and absence seizures (Blumenfeld, 2005)—conditions that are allegedly characterized by high levels of arousal but in which consciousness is either altogether absent or, if present, present only in terms of modes with highly restricted contents and impaired structure and access.

The second kind of problem case for the arousal-based account derives from conditions in which subjects are low in arousal but their mode of consciousness is not uniformly low. A key example of such a state is REM dreaming. Arousal (at least insofar as this notion is understood behaviorally) is low in

REM dreaming, but subjects enjoy (or at least appear to enjoy) many conscious experiences with few limits on content, with some degree of structure, and often with a semblance of accessibility (Windt, 2011). Of course, there may be respects in which REM dreaming involves more compromised modes of consciousness in terms of content, structure, and access. In particular, the functional systems responsible for reasoning and decision making appear to be off-line, so perhaps what REM dreaming really shows is that the content-based aspects of modes can dissociate from the functional-based aspects.

Although the picture is not wholly clear, these complications put significant pressure on the idea that a single mechanism—the "arousal mechanism"—simultaneously upregulates or downregulates the various systems that determines a creature's mode of consciousness. A more complicated mechanism is needed, one that can affect different aspects of the mode of consciousness separately but, nevertheless, in such a way that it makes sense to talk of a global mode with degrees of change. Although it is tempting to see arousal as the common cause of changes in the mode of consciousness, the case is far from watertight: Not only is it somewhat unclear what exactly arousal is, different modes seem to be differentially influenced by the subject's level of arousal.

It is sometimes suggested that "wakefulness" rather than "arousal" might unify the various aspects of modes that we have identified. However, this approach inherits many of the problems that confront the arousal-based proposal. For one thing, it is not always entirely clear that the notion of wakefulness is entirely distinct from that of arousal. For example, whereas Laureys (2005) labeled the horizontal axis of Figure 5.1 "wakefulness," Mormann and Koch (2007) labeled the horizontal axis of their almost identical figure "level of arousal." Zeman (2001, p. 1265) says that "to be conscious is to be awake, aroused, alert, or vigilant," whereas others define wakefulness in terms of "arousal/responsiveness, circadian rhythms, sleep cycle, and homeostasis" (Bekinschtein et al., 2009, p. 172). Moreover, the behavioral capacities that associate with wakefulness can fractionate, in the sense that certain aspects of wakefulness can be present while others are absent. As Zeman (2001) noted,

> "Wakefulness", after all, is shorthand for a set of associated neural, behavioural and psychological functions which are, to some extent, independently controlled. This is evident from a range of pathological states in which the usual associations between these states break down. Thus, for example, a sleepwalker is capable of coordinated movement, may be able to avoid obstacles and accede to gentle urging to return to bed, but will later

> have no recall of the episode; motor control, perception and memory have lost their usual relationship. (p. 1271)

If wakefulness fractionates in this manner, then it becomes harder to tell a story in which wakefulness is the common cause of straightforwardly ordered co-variation in the three aspects of modes. Moreover, relatively high levels of wakefulness can coincide with modes of consciousness that have severe content limits and compromised structure and access.

CONCLUDING REMARKS

We began by observing that the notion of a mode of consciousness is best grasped by means of examples as opposed to a formal definition. We will bring this chapter to a close by considering how the model of modes sketched here might bear on our understanding of disorders of consciousness—in particular, the vegetative state and the MCS.

Perhaps the most fundamental question posed by these disorders is whether patients in these states are indeed (ever) conscious. That is a question on which our account of modes is officially silent. Nevertheless, our account might—in principle, at least—enable us to answer questions about how many modes of consciousness are associated with these states and how those modes are related to each other.

One possibility is that the vegetative state and the MCS are associated with two modes of consciousness that are clearly distinguishable according to each of the three dimensions of modes that we have examined here. (Of course, most patients in a vegetative state are presumably not conscious at all; here we refer only to that small percentage of patients in a behaviorally vegetative state who are conscious.) According to this picture, conscious patients in a vegetative state would be distinguishable from conscious MCS patients on the basis of (1) the range of contents that are consciously accessible to them; (2) the temporal and attentional structure of their experience; and (3) the degree to which their conscious contents are able to drive cognitive and behavioral control. Although this possibility could be correct, we suspect that the reality will prove to be significantly messier. In fact, we suspect that this tidy picture will break down in at least two ways.

First, the three modal dimensions of consciousness might turn out to dissociate from each other, such that certain patients might, for example, turn out be in a vegetative state according to the content-related dimension but in MCS according to the functional dimension. (This possibility might not be realized if the three modal dimensions of consciousness are directly underpinned by a single mechanism, but we have already noted that that proposal is an open one at present.) Second, it might turn out that even if the three modal dimensions of consciousness are tightly correlated with each other, there is no bright dividing line between the global states of consciousness associated with patients who are clinically in a vegetative state and those who are clinically in a MCS. In other words, it might turn out that as far as consciousness is concerned, we should really think in terms of overlapping regions within a multidimensional space. Some patients in a vegetative state will be located in one region of this apace, and some MCS patients will be located in another region, but there may be no clear boundaries within this space that distinguish the 'VS' mode of consciousness from the 'MCS' mode of consciousness. Instead, we might simply have to recognize a series of modes that are "increasingly MCS" in nature. Of course, the foregoing is mere speculation. Our point here is not to argue in favor of one or another taxonomy of disorders of consciousness but to simply illustrate how the account of modes that we have developed here might illuminate a set of issues that remain somewhat obscure at present.[9]

NOTES

1. What we are here calling modes of consciousness are often referred to by clinicians and neuroscientists as "states of consciousness." We use the former term on the grounds that modes are merely one kind of state of consciousness alongside the more familiar fine-grained states of consciousness that are individuated by appeal to their contents.
2. We are much indebted to discussions with Lisa Miracchi on this point.
3. Chalmers' use of the label "creature consciousness" is very different from David Rosenthal's (1986) use of the term; Rosenthal uses it as a synonym for "wakefulness."
4. A variant of the bandwidth approach focuses not on the current number of items of consciousness but on the temporal dynamics of the information channel (e.g., how swiftly the items in consciousness can change in response to external and internal stimuli). For example, in some modes, the contents of consciousness might shift only very sluggishly. We consider an approach of this kind in more detail later in this chapter.
5. Although talk of levels of consciousness is occasionally encountered in developmental psychology and cognitive ethology, the notion of levels that is employed in these disciplines is distinct from that which is employed in clinical neurology. Exactly how to characterize this contrast is not straightforward, but roughly speaking, the levels of consciousness that are invoked in the former disciplines are generally fixed by relatively stable features of a creature's cognitive and perceptual architecture, whereas the levels of consciousness that are invoked by clinical neurology are generally independent of such features and subjects may transition from one "level" to another in a manner of hours or even minutes.
6. It is controversial whether "unattended" contents are completely unattended or merely less attended than contents that fall within the attentional focus, but for ease of discussion we will refer to them as unattended.

7. Some aspects of attention may belong with a notion of content insofar as changes in attentional focus might track changes in the fineness of representational grain (Block, 2010; Stazicker, 2011), but it is far from clear that the notion of attentional focus can be fully captured by appeal to content-based factors. In fact, the core notion of attention appears to be functional: What it is for contents to fall within the focus of attention is for them to have preferential access to high-level processing. Versions of this idea can be seen in accounts that treat attention in terms of biased competition (Desimone & Duncan, 1995) or precision optimization (Feldman & Friston, 2010; Rao, 2005). (We turn to functional approaches to modes in the next section).
8. It is also possible that there is a hierarchical relation between consuming systems, such that some consuming systems can be switched on only if other systems are also switched on. In this respect, it is interesting that studies of preserved functioning in the context of epileptic absence seizures indicate that certain types of tasks (e.g., choice reaction tasks, short-term memory tasks) are more likely to be preserved than is the ability to count or read aloud, and both of these capacities are more likely to be preserved than is the ability to respond to commands or verbal questions (Blumenfeld, 2005).
9. This chapter was written with the support of European Research Council grant 313552, *The Architecture of Consciousness,* and Australian Research Council grant DP0984572, *Conscious States in Conscious Creatures: A Philosophical Framework for the Science of Consciousness.* We are very grateful to Tom McClelland, Lisa Miracchi, Walter Sinnott-Armstrong, and Adam Zeman for comments on a previous version.

REFERENCES

Alkire, M. T., & Miller, J. (2005). General anesthesia and the neural correlates of consciousness. *Progress in Brain Research, 150*: 229–244.

Allman, M. H., & Meck, W. H. (2012). Pathophysiological distortions in time perception and timed performance. *Brain, 135*, 656–677.

Bekinschtein, T. A., Dehaene, S., Rohaut, B., Tadel, F., Cohen, &Naccache, L. (2009). Neural signature of the conscious processing of auditory regularities. *Proceedings of the National Academy of Sciences U S A, 106*(5), 1672–1677.

Bhat, R., & Rockwood, K. (2007). Delirium as a disorder of consciousness. *Journal of Neurology, Neurosurgery and Psychiatry, 78*, 1167–1170.

Block, N. (2010). Attention and mental paint. *Philosophical Issues, 20*, 23–63.

Blumenfeld, H. (2005). Consciousness and epilepsy: Why are patients with absence seizures absent? In S. Laureys (Ed.). *Progress in Brain Research* (Vol. 150): *The Boundaries of Consciousness* (pp. 271–287). Amsterdam: Elsevier.

Boly, M., Phillips, C., Balteau, E., Schnakers, C., Degueldre, C., Moonen, G., . . . & Laureys, S. (2008). Consciousness and cerebral baseline activity fluctuations. *Human Brain Mapping, 29*(7), 868–874.

Buckner, R. L., Andrews-Hanna, J. R., & Schacter, D. L. (2008). The brain's default network: Anatomy, function, and relevance to disease. *Annals of the New York Academy of Sciences, 1124*(1), 1–38.

Chalmers, J. (1998). What is a neural correlate of consciousness? In T. Metzinger (Ed.). *Neural correlates of consciousness: Empirical and conceptual questions.* Cambridge, MA: MIT Press.

Corlett, P. R., Honey, G.D., & Fletcher, P. C. (2007). From prediction error to psychosis: Ketamine as a pharmacological model of delusions. *Journal of Psychopharmacology, 21,* 238–252.

Cowan, N. (2000). The magical number 4 in short-term memory: A reconsideration of short-term mental storage capacity. *Behavioral and Brain Sciences, 24,* 87–185.

Demertzi, A., Vanhaudenhuyse, A., Bruno, M. A., Schnakers, C., Boly, M., Boveroux, P., . . . & Laureys, S. (2008). Is there anybody in there? Detecting awareness in disorders of consciousness. *Expert Review of Neurotherapeutics, 8*(11), 1719–1730.

Desimone, R., & Duncan, J. (1995). Neural mechanisms of selective visual attention. *Annual Review of Neuroscience, 18,* 193–222.

Durmer, J. S., & Dinges, D. F. (2005). Neurocognitive consequences of sleep deprivation. *Seminars in Neurology, 25*(1), 117–129.

Englot, D. J., & Blumenfeld, H. (2009). Consciousness and epilepsy: Why are complex-partial seizures complex? *Progress in Brain Research, 177,* 147–170.

Feldman, H., & Friston, K. (2010). Attention, uncertainty, and free-energy. *Frontiers in Human Neuroscience, 4,* 215.

Frackowiak, R. S. Friston, K. J., Frith, C. D., Dolan, R. J., Price, C. J., Zeki, S., . . . & Penny, W. D. (Eds.). (2004). *Human brain function.* London, Academic Press.

Geyer, M. A., & Vollenweider, F. X. (2008). Serotonin research: Contributions to understanding psychoses. *Trends in Pharmacological Science, 29,* 445–453.

Giacino J. T., Kalmar K., & Whyte, J. (2004). The JFK Coma Recovery Scale-Revised: Measurement characteristics and diagnostic utility. *Archives of Physical Medicine and Rehabilitation, 85,* 2020–2029.

Giacino, J. T., Schnakers, C., Rodriguez-Moreno, D., Kalmar, K., Schiff, N., & Hirsch, J. (2009). Behavioral assessment in patients with disorders of consciousness: Gold standard or fool's gold? *Progress in Brain Research, 177,* 33–48.

Greicius, M. D., Krasnow, B., Reiss, A. L., & Menon, V. (2003). Functional connectivity in the resting brain: A network analysis of the default mode hypothesis. *Proceedings of the National Academy of Sciences USA, 100,* 253–258.

Grondin, S. (2010). Timing and time perception: A review of recent behavioral and neuroscience findings and theoretical directions. *Attention, Perception, and Psychophysics, 72*(3), 561-582.

Gusnard, D. A. (2005). Being a self: Considerations from functional imaging. *Consciousness and Cognition, 14*(4), 679.

Hobson, J. A. (1999). *Dreaming as delirium: How the brain goes out of its mind.* Cambridge, MA: MIT Press.

Jamieson, G. (Ed.). (2007). *Hypnosis and conscious states.* Oxford: Oxford University Press.

Johanson, M., Revonsuo, A., Chaplin, J., & Wedlund, J.-E. (2003). Level and contents of consciousness in connection with partial epileptic seizures. *Epilepsy and Behavior, 4,* 279–285.

Kriegel, U. (2004). The functional role of consciousness: A phenomenological approach. *Phenomenology and the Cognitive Sciences, 3*(2), 171–193.

Laureys, S. (2005). The neural correlate of (un)awareness: Lessons from the vegetative state. *Trends in Cognitive Sciences, 9*(12), 556-559.

Mahowald, M. W., & Schenck, C. H. (1992). Dissociated states of wakefulness and sleep. *Neurology, 42*(Suppl 6), 44–52.

Mormann, F., & Koch, C. (2007). Neural correlates of consciousness. *Scholarpedia, 2*(12), 1740.

Owen, A. M. (2008). Disorders of consciousness. *Annals of the New York Academy of Sciences, 1124*(1), 225–238.

Panksepp, J. (2005). Affective consciousness: Core emotional feelings in animals and humans. *Conscious and Cognition, 14*, 30–80.

Panksepp, J., Fuchs, T., Garcia, V., & Lesiak, A. (2007). Does any aspect of mind survive brain damage that typically leads to a persistent vegetative state? Ethical considerations. *Philosophy, Ethics, and Humanities in Medicine, 2*(1), 32.

Pockett, S. (2003). How long is "now"? Phenomenology and the specious present. *Phenomenology and the Cognitive Sciences, 2*(1), 55–68.

Posner, J. B., Saper, C. B., Schiff, N., & Plum, F. (2007). *Plum and Posner's diagnosis of stupor and coma* (4th ed.). Oxford: Oxford University Press.

Rao, R. P. N. (2005). Bayesian inference and attentional modulation in the visual cortex. *NeuroReport, 16*(16), 1843–1848.

Rosenthal, D. (1986). Two concepts of consciousness. *Philosophical Studies, 94*, 329–359.

Schiff, N. D., & Plum, F. (2000). The role of arousal and "gating" systems in the neurology of impaired consciousness. *Journal of Clinical Neurophysiology, 17*(5), 438–452.

Stazicker, J. (2011). Attention, visual consciousness and indeterminacy. *Mind and Language, 26*(2), 156–184.

Watzl, S. (2012). Attention as structuring of the stream of consciousness. In C. Mole, D. Smithies, and W. Wu (Eds.) *Attention: Philosophical and psychological essays.* Oxford: Oxford University Press.

Windt, J. (2011). Altered consciousness in philosophy. In E. Cardena & M. Winkelman (Eds.) *Altering consciousness: Multidisciplinary perspectives.* Santa Barbara: CA: ABC-CLIO.

Zeman, A. (2001). Consciousness. *Brain, 124*, 1263-1289.

PART II

Diagnosis

6

What Is It Like To Be in a Disorder of Consciousness?

CAROLINE SCHNAKERS

I'm the only one who can know directly that I am conscious. Nevertheless, I can interact with others using verbal communication or motor responses to let them know that I am conscious. What would happen if, to some extent, I knew I was conscious but was unable to express it? How would people around me know? The challenge facing clinicians involved in the care of patients recovering from coma lies in differentiating reflex from voluntary activity—in other words, detecting consciousness in noncommunicative patients (Schnakers, Laureys, & Majerus, 2011).

Sixty years ago, the field of disorders of consciousness was a very limited research domain. Severely brain-injured patients, who are most likely to exhibit impaired consciousness during recovery, often died. In the 1950s, the introduction of artificial breathing changed everything. The lives of these patients could be extended even with severe lesions to brain areas supporting the control of vital functions. Clinicians began to face patients who were alive but not reactive to their surroundings. A new field emerged: the disorders of consciousness. This review presents the main clinical entities that severely brain-injured patients go through before fully recovering from coma and explains how to assess consciousness and what brain activity is like in conscious versus unconscious patients.

DISORDERS OF CONSCIOUSNESS

Coma

Patients who survive a severe brain injury can remain unconscious for several weeks, being neither awake nor conscious. They are in a state called *coma*, defined as "a pathological state marked by severe and prolonged dysfunction of vigilance and consciousness" (Posner, Saper, Schiff, & Plum, 2007) (Figure 6.1). This state usually results either from a lesion limited to the brainstem (involving the reticular activating system) or from a global brain dysfunction (most often caused by diffuse axonal injury after traumatic brain injury). The distinguishing features of coma are the continuous absence of eye opening (spontaneously or after stimulation) and the absence of oriented or voluntary motor or verbal responses. The prognosis is influenced by factors such as etiology (traumatic patients have a better outcome), general medical condition, and age. A bad outcome is expected if, after 3 days of observation, there are no pupillary or corneal reflexes, only stereotyped or absent motor responses to noxious stimulation, and an isoelectric electroencephalograph (EEG) pattern. A bilateral absence of the parietal N20 peak when somatosensory evoked potentials are recorded is also a strong predictor of death in comatose patients (Bouwes et al., 2012).

Being in a coma is different from being brain dead. Brain death suggests that the organism cannot function as a whole. Critical functions such as respiration and circulation, neuroendocrine and homeostatic regulation, and consciousness are permanently absent. The patient is apneic and unreactive to environmental stimulation. The term *brain death* requires the bedside demonstration of irreversible cessation of all clinical functions of the brain, including the brainstem. After excluding the impact of pharmacological

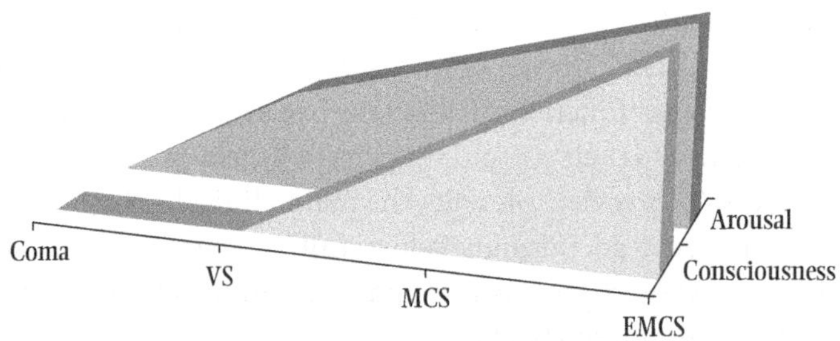

Figure 6.1 The main clinical entities encountered when recovering from coma, illustrated as a function of two main components: arousal and consciousness. EMCS = emergence from MCS; MCS = minimally conscious state; VS = vegetative state.

treatments, toxins, and hypothermia, the diagnosis can be made within 6 to 24 hours after injury (Laureys, 2005). This condition is permanent, in contrast to coma, although comatose patients may progress to brain death. In cases with a better outcome, patients may emerge from coma within 2 to 4 weeks, evolving into a vegetative state or a minimally conscious state (Posner et al., 2007).

Vegetative State

In 1972, the term *vegetative state* (VS) was first introduced by Jennet and Plum to describe "an organic body capable of growth and development but devoid of sensation and thought" (Jennett & Plum, 1972). This new clinical entity was identified after the development and use of artificial breathing techniques in intensive care units. Since the definition of this state, the number of scientific studies performed on VS patients has continuously increased. More precisely, only 15 articles were published between 1975 to 1985, compared with 393 articles from 2000 to 2010 (Figure 6.2).

This interest is comprehensible because the VS state is more than astonishing. Behaviorally, patients in VS open their eyes spontaneously or in

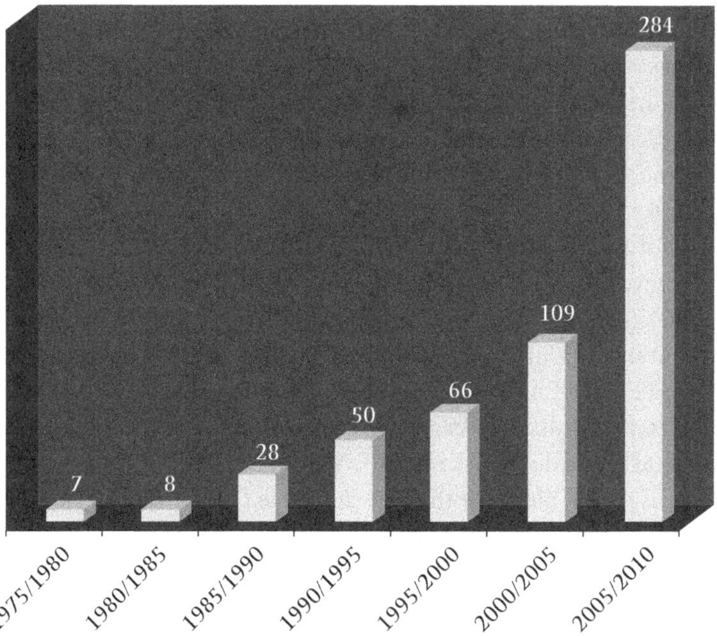

Figure 6.2 Annual number of publications on vegetative state between 1975 and 2010. Results of a PubMed search using the terms vegetative state, consciousness, and awareness.

response to stimulation and present preserved autonomic functions (e.g., cardiovascular regulation, thermoregulation), but they are not conscious and show only reflex behaviors (Multi-Society Task Force, 1994). The VS often results from an injury involving the white matter or bilateral lesions of the thalamus (i.e., intralaminar nuclei) (Fernández-Espejo et al., 2010; Newcombe et al., 2010). How can doctors explain to the family such a state? These patients breathe normally and have their eyes open. They even have prolonged periods with eyes closed, miming sleep. We nevertheless know now that this appearance is misleading. When the eyes are closed, no electroencephalographical change is noticed, and no common stage of sleep (e.g., slow wave sleep, rapid eye movement) is present (Landsness et al., 2011). The patients may also moan and show smiling, crying, or grimacing even though these behaviors are inappropriate and appear out of context. All of these features may puzzle the family and complicate the work of the medical staff, who are may be inclined to experience burnout (Gosseries et al., 2012).

Accurate information and psychological assistance are essential and may help the family (and hence the medical staff) to cope with this dramatic situation (Leonardi et al., 2012). The VS state may last for days, months, or even years. After 1 year for traumatic etiologies or 3 months for nontraumatic etiologies, the VS can be considered as permanent. In such cases, the patient's chance of recovery is less than 5%. Only after that point can the ethical and legal issues concerning withdrawal of treatment be discussed (Jennett, 2005). Given the negative connotation of the term "vegetative state," the European Task Force on Disorders of Consciousness has proposed use of the more neutral and descriptive term, *unresponsive wakefulness syndrome* (Laureys et al., 2010).

In end-of-life decision making, the presence of any sign of consciousness must be investigated and excluded using behavioral but also paramedical approaches (see later discussion). The most famous cases have been those of Tony Blant in the United Kingdom (1993), Terri Schiavo in the United States (2005), and Eluana Englaro in Italy (2009) (Andrews, 2004; Cohen & Kummer, 2006; Luchetti, 2010). Deep examinations of all those patients were performed before the extreme decision to end life was considered.

These cases also illustrate the real challenge that clinicians face: the differentiation between unconscious and conscious states. This distinction is not as easy to determine as it may seem. Although the patients are showing no obvious conscious behaviors and perhaps no responses at all, they may yet present subtle signs of consciousness that clinicians must detect. Such patients are not considered as being fully conscious; they are in what is termed a *minimally conscious state* (MCS).

Minimally Conscious State

The MCS clinical entity was identified more recently than the VS and the coma. It was defined in 2002 by the Aspen Workgroup as being characterized by the presence of inconsistent but clearly discernible behavioral signs of consciousness (see Figure 6.1) (Giacino et al., 2002). Patients evolving from a VS to a MCS are still awake but begin to show oriented behaviors (e.g., visual pursuit). The earlier this behavior appears, better the outcome will be. The presence of oriented eye movement is therefore crucial, but it is also one of the most difficult signs of consciousness to detect and requires the use of sensitive diagnostic tools (see later discussion) (Candelieri et al., 2011; Dolce et al., 2011; Schnakers et al., 2009). More globally, signs of consciousness in MCS patients may be hard to observe because they are inconsistent in time due to high vigilance fluctuations. They nevertheless must be replicated within a given examination to meet the diagnostic criteria for MCS.

Later, MCS patients may also start to understand language and respond to simple commands (e.g., "Shake my hand"). As different functional neuroanatomical features have been observed, the MCS has recently been subdivided in two clinical entities, MCS+ and MCS−, based on the presence or absence, respectively, of responses to command (Bruno et al., 2012b). Patients in an MCS may also show appropriate emotional responses (e.g., crying or laughing in contingent relation to appropriate environmental triggers), inappropriate object manipulation, or intelligible verbalizations without being able to functionally communicate either verbally or by gestures.

Recovery from MCS is defined by the re-emergence of functional communication and/or functional use of objects Giacino et al., 2002). Some patients remain in an MCS without fully recovering consciousness for an extended period. Even though the probability of functional recovery at 1 year is more favorable relative to VS for either traumatic or nontraumatic etiologies (Bruno, 2012a), prognosis remains very difficult because of the great heterogeneity of recovery in chronic patients. Some patients in MCS progress slowly, whereas others remain in this condition permanently (Fins, Schiff, & Foley, 2007).

HOW CONSCIOUSNESS IS DETECTED

Until now, the main way to know whether a patient is conscious has been by observing what the patient can do spontaneously or in response to stimulation. When going to the patient's bedside, the clinician must keep in mind two major components: arousal and consciousness. Clinically, arousal is manifested by spontaneous eye opening. Consciousness, on the other hand, can be assessed by oriented or voluntary responses to external stimuli (e.g., visual pursuit, command following). Consciousness requires a certain level of arousal, but even a preserved level of arousal does not necessarily imply consciousness (e.g., in the VS).

Behavioral assessment requires thorough expertise. Differentiating MCS from VS can be challenging because voluntary and reflex behaviors may be difficult to distinguish, and subtle signs of consciousness may be missed. Studies have shown that approximately 40% of patients diagnosed as being in VS were misdiagnosed and were in fact conscious (Andrews, Murphy, Munday, & Littlewood, 1996; Childs, Mercer, & Childs, 1993; Schnakers et al., 2009). This high misdiagnosis rate likely reflects different sources of variance. Variance in diagnostic accuracy may result from biases contributed by the examiner, the patient, and the environment (Gill-Thwaites, 2006).

Examiner error may arise when the range of behaviors sampled is too narrow, response-time windows are overinclusive or underinclusive, criteria for judging purposeful responses are poorly defined, or examinations are conducted too infrequently to capture the full range of behavioral fluctuation. The use of standardized rating scales offers some protection from these errors, although failure to adhere to specific administration and scoring guidelines may jeopardize diagnostic accuracy. The second source of variance concerns the patient. Fluctuations in arousal level, fatigue, subclinical seizure activity, occult illness, pain, cortical sensory deficits (e.g., cortical blindness/deafness), motor impairment (e.g., generalized hypotonus, spasticity, paralysis), or cognitive disturbance (e.g., aphasia, apraxia, agnosia) constitute a bias to the behavioral assessment and therefore decrease the probability of observing signs of consciousness. Finally, the environment in which the patient is evaluated may bias assessment findings. Paralytic and sedating medications, restricted range of movement stemming from restraints and immobilization techniques, poor positioning, and excessive ambient noise, heat, or light can decrease or distort voluntary behavioral responses.

Some sources of error can be avoided, but this is not always possible or under the examiner's control. Nevertheless, eliminating error is particularly important because clinical management, from treatment of pain to end-of-life decision making, often depends on behavioral observations. Developing valid and sensitive behavioral scales helps clinicians to detect the presence of signs of consciousness, even subtle ones, and therefore represents a real necessity.

The Glasgow Coma Scale

The Glasgow Coma Scale (GSC) was the first behavioral scale to be widely used. This scale was the first validated rating scale developed to monitor levels of consciousness in intensive care units (Teasdale & Jennett, 1976). It includes three subscales that address arousal level, motor function, and verbal abilities. The

GCS has been extensively investigated, particularly for its prognostic value. Nevertheless, despite its widespread use, the GCS has been criticized for its lack of inter-rater agreement and issues regarding its use in patients with ocular trauma, tracheostomy, or ventilatory support (McNett, 2007). The use of this scale may also lead to misdiagnosis because the GCS does not include signs of consciousness such as object manipulation, appropriate emotional responses, or visual pursuit—which, as explained earlier, provide crucial prognostic value. Moreover, one study (Schnakers et al., 2006) showed that use of such a scale could lead to a misdiagnosis rate of 50%, suggesting that it is not sufficient for a scale to be standardized; it also must be a sensitive diagnostic tool.

The Coma Recovery Scale

The Coma Recovery Scale-Revised (CRS-R) aims to improve differential diagnosis in patients with disorders of consciousness (Figure 6.3) (Giacino, Kalmar, & Whyte, 2004). The scale consists of twenty-three items that comprise six subscales addressing auditory, visual, motor, oromotor, communication, and arousal functions. CRS-R subscales are composed of hierarchically arranged items associated with brainstem, subcortical, and cortical processes. The lowest item on each subscale represents reflex activity, whereas the highest items represent cognitively mediated behaviors. Scoring is standardized and is based on the presence or absence of operationally defined behavioral responses to specific sensory stimuli. Psychometric studies indicate that the CRS-R meets high standards for measurement and evaluation tools designed for use in interdisciplinary medical rehabilitation. The CRS-R can be administered reliably by trained examiners and produces reasonably stable scores over repeated assessments. Validity analyses have shown that the CRS-R is capable of distinguishing between patients in MCS and those in VS even better than other consciousness scales (e.g., GCS), and this distinction is of critical importance in establishing prognosis and formulating treatment interventions (Schnakers et al., 2006; Seel et al., 2010).

Even though the CRS-R constitutes a very sensitive diagnostic tool, it involves various components linked to cognition and perception. It does not give precise information on each of those components. The development of tools providing such information would further help clinicians to refine rehabilitative interventions and would certainly have an impact on patients' recovery.

The Nociception Coma Scale

"Is my loved one in pain?" is one of the first questions the family wonders when facing a relative in a disorder of consciousness. The clinical expert

AUDITORY FUNCTION SCALE

4 - Consistent Movement to Command[*]
3 - Reproducible Movement to Command[*]
2 - Localization to Sound
1 - Auditory Startle
0 - None

VISUAL FUNCTION SCALE

5 - Object Recognition[*]
4 - Object Localization: Reaching[*]
3 - Pursuit Eye Movements[*]
2 - Fixation[*]
1 - Visual Startle
0 - None

MOTOR FUNCTION SCALE

6 - Functional Object Use[t]
5 - Automatic Motor Response[*]
4 - Object Manipulation[*]
3 - Localization to Noxious Stimulation[*]
2 - Flexion Withdrawal
1 - Abnormal Posturing
0 - None/Flaccid

OROMOTOR/VERBAL FUNCTION SCALE

3 - Intelligible Verbalization[*]
2 - Vocalization/Oral Movement
1 - Oral Reflexive Movement
0 - None

COMMUNICATION SCALE

3 - Oriented[t]
2 - Functional: Accurate[t]
1 - Non-Functional: Intentional[*]
0 - None

AROUSAL SCALE

3 - Attention[*]
2 - Eye Opening w/o Stimulation
1 - Eye Opening with Stimulation
0 - Unarousable

[*] indicates minimally conscious state
[t] indicates emergence form minimally conscious state

Figure 6.3 Protocol of the Coma Recovery Scale-Revised (CRS-R).

might explain what was found in previous neuroimaging studies (see later discussion). However, he or she will be puzzled when asked about pain monitoring. Until recently, no tool allowed monitoring of such a component in severely brain-injured patients. In order to respond this need, a new scale was developed, the Nociceptive Coma Scale (NCS) (Schnakers et al., 2010). The NCS was based on preexisting pain scales that were developed for noncommunicative patients with advanced dementia or for newborns (Schnakers et al., 2012).

The NCS consists of four subscales that assess motor, verbal, and visual responses and facial expression. The NCS has been validated in patients from intensive care units, neurology/neurosurgery units, rehabilitation centers, and nursing homes. The scale demonstrates good inter-rater reliability and good concurrent validity. Compared with other pain scales developed for noncommunicative patients, the NCS has shown a broader score range and a better sensitivity to clinical diagnosis (i.e., VS versus MCS). The motor, verbal, and facial subscores included in the revised version of the scale (NCS-R) were also found to be significantly higher in response to noxious stimuli than non-noxious stimuli, reflecting the good sensitivity of the scale. Based on the NCS-R, a cutoff score of 4 (which has a sensitivity of 73% and a specificity of 97%) has been defined as a potential clinical threshold for detection of pain in patients with disorders of consciousness (Chatelle, 2012).

RESIDUAL BRAIN ACTIVITY

As discussed earlier, differentiating conscious from unconscious patients is particularly challenging. Since the late 1990s, the use of neuroimaging techniques has improved our understanding of disorders of consciousness, allowing better differentiation between these populations and, more particularly, between the residual brain activity of VS versus MCS patients (Figure 6.4).

In VS patients, brain hypometabolism is similar to that of patients in coma, who have a 50% to 60% decreased global metabolic rate relative to controls. Hypometabolic activity is further reduced to 60% to 70% in patients in permanent VS (Tommasino et al., 1995). However, the global metabolic rate does not always come back to normal after recovery, and sometimes it does not show any substantial changes (Laureys, 1999b). Global brain metabolism therefore does not reflect the presence or absence of consciousness.

Another potential approach is to compare brain areas and determine whether some areas are more important than others. In patients in VS, metabolic dysfunctions have been reported in wide parts of the brain, including the polymodal associative cortices: lateral and medial frontal regions bilaterally, parietotemporal and posterior parietal areas bilaterally, posterior

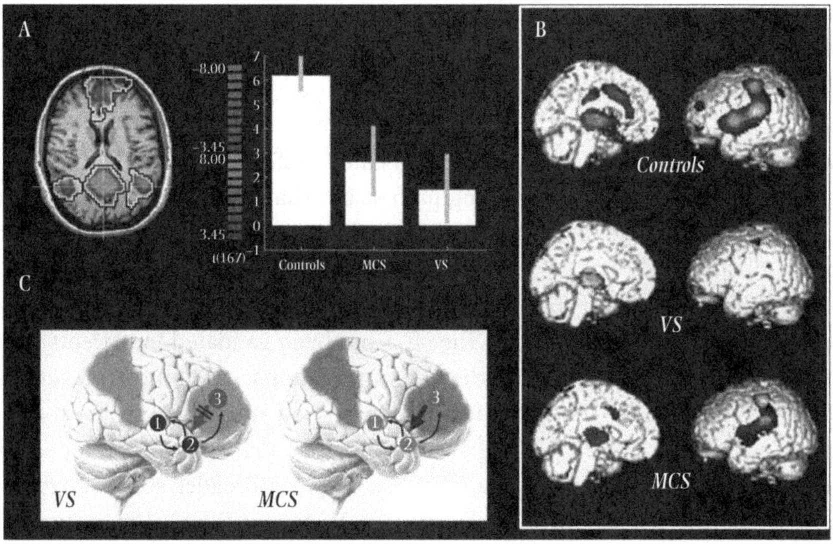

Figure 6.4 Residual brain activity in disorders of consciousness. Residual brain activity at rest (within the default network) (A) and in response to noxious stimulation (B) in patients in a vegetative state (VS) compared with patients in a minimally conscious state (MCS). (A and B adapted from Schnakers et al., 2012, and Vanhaudenhuyse et al., 2010). Panel C illustrates the deficit in connectivity (particularly, long distance backward connectivity) in VS compared with MCS patients (adapted from Boly et al., 2011). 1 = Primary Auditory Cortex; 2 = Superior Temporal Gyrus; 3 = Inferior Frontal Gyrus.

cingulate, and precuneal cortices (Laureys et al., 1999a). These impaired areas are involved in the default network, which is mostly active at rest and which is implicated in cognitive processes such as daydreaming or mind-wandering, stimulus-independent thoughts, and self-related thoughts (Anticevic et al., 2012). In addition to metabolic dysfunctions in these areas, studies have shown that connectivity within the default network is lower in VS patients compared with conscious patients (Cauda et al., 2009; Crone et al., 2011; Fernández-Espejo et al., 2012; Vanhaudenhuyse et al., 2010). More specifically, the precuneus (an area known to be crucial in conscious processing) was found to be significantly less connected to the other areas of the network in VS patients compared with MCS patients (Vanhaudenhuyse et al., 2010).

Similar findings have been found in studies investigating brain activity in response to stimulation. In VS patients, activation studies using auditory stimulation (i.e., tones) showed preserved functioning in the primary auditory cortex without, however, encompassing other brain areas (e.g., the temporoparietal junction) (Laureys et al., 2000a). Similarly, in VS patients, noxious stimulation (i.e., electrical stimulation of the median nerve) activated only a part of the network involved in the low-level sensory-discriminative processing of pain (i.e., midbrain, contralateral thalamus, and primary somatosensory cortex) (Laureys et al., 2002). In both activation studies, low-level primary cortical activity seemed to be isolated from higher-level associative cortical activity (Laureys et al., 2000a; 2002). More exactly, studies suggest that long-distance connectivity (e.g., between frontal to temporal areas) is more affected than short-distance connectivity (e.g., areas within the temporal gyrus), which may be crucial for integrative brain processing leading to consciousness (Boly et al., 2011).

There is disruption not only in the way that the information is processed at a cortical level but also in the way that it is transmitted from a subcortical to a cortical level. The ascending reticular activating system seems disconnected from posterior medial cortices (i.e., precuneus) (Silva et al., 2010); more importantly, a re-establishment of the connections between the thalamus and associative areas has been found in patients who recovered from the VS (Laureys et al., 2000b). This is particularly important because the thalamus constitutes a relay in transmitting sensory and motor signals from subcortical to cortical areas and has a key role in conscious processing (Ward, 2011). In fact, several studies have reported functional and structural impairments in this area in VS (Fernández-Espejo et al., 2010; Lull et al., 2010).

These results suggest the presence of an impaired and disconnected residual brain activity in VS patients which, most likely, does not lead to integrated conscious perception. This differs from the situation in MCS patients. In the latter group, despite a total metabolic depression of 20% to 40% (Schiff et al., 2005), the activity in precuneus and posterior cingulate cortex (the most active regions in awakening and the least active ones under general anesthesia or during deep slow wave sleep) is superior to that observed in VS patients (Laureys et al., 2005). In contrast with the limited brain activation found in VS, functional imaging studies using auditory stimulation showed a larger temporal activation (encompassing the temporoparietal junction) (Bekinschtein et al., 2004; Boly et al., 2004; Schiff et al., 2005). The valence of an auditory stimulation can lead to a difference of brain activation in these patients. Indeed, MCS patients seem to activate a broader part of the temporal lobe and the amygdala in response to an emotional auditory stimulation than in response to a nonemotional auditory stimulation (Bekinschtein et al., 2004).

For other types of stimulation (e.g., noxious stimulation), Boly et al. showed brain activation similar to that observed in controls (involving the anterior cingulate area, which is known to be linked to pain unpleasantness) (Boly et al., 2005; 2008). In parallel, a longitudinal case report described by Bekinschtein et al. showed a fronto-temporo-parietal activation, not just an isolated temporal left activation, in response to the presentation of words when patients evolved from a VS toward an MCS (Bekinschtein et al., 2005). Higher activation in the thalamus has also been observed in MCS patients compared with VS patients (Fernández-Espejo, 2011). Finally, in addition to higher brain activation, a higher cortical and thalamocortical connectivity has been noticed in MCS patients compared with VS patients (Boly et al., 2004; 2008; Laureys et al., 2000b; Rosanova et al., 2012). These data suggest that MCS patients may have sufficient cortical integration and access to afferent information to allow conscious perception.

CONCLUSION

Behavioral responses and brain activity differ among the various disorders of consciousness. The detection at the bedside of signs of consciousness can be challenging, and the use of sensitive standardized scales is crucial. Because misdiagnosis can lead to serious consequences, especially in terms of pain treatment and end-of-life decision making, neuroimaging could constitute a complementary tool to distinguish VS from MCS. In the future, the development of a consciousness classifier based on residual brain activity (Phillips et al., 2011) or residual brain connectivity (Casali et al., 2013) could also substantially help clinicians and constitute an automated diagnostic tool. This could be particularly helpful for detection of patients who are functionally locked-in (Bruno et al., 2011).

A series of neuroimaging and electrophysiological studies using various active tasks (e.g., mental imagery, counting) have shown voluntary responses in patients who remained behaviorally unresponsive (Cruse et al., 2011; Faugeras et al., 2011; Monti et al., 2010; Owen & Coleman, 2008; Schnakers et al., 2008). A large cohort study found such responses in 10% of patients with disorders of consciousness and functional communication in one patient (2%) (Monti et al., 2010). The discovery of this new clinical entity requires us to consider its ethical implications. Indeed, how do we know that a patient for whom discontinuation of life support measures is being considered is truly in a VS? The demographic characteristics as well as the cognitive profile of such an entity have to be investigated as soon as possible. However, the active tasks used currently request high-level cognitive functioning and may underestimate the percentage of severely brain-injured patients who are conscious but behaviorally unresponsive. The development of a passive paradigm that would allow the detection of consciousness in those patients based on residual brain activity or residual brain connectivity would therefore constitute a tremendous step forward for the field.

REFERENCES

Andrews, K. (2004). Medical decision making in the vegetative state: Withdrawal of nutrition and hydration. *NeuroRehabilitation, 19*, 299–304.

Andrews, K., Murphy, L., Munday, R., & Littlewood, C. (1996). Misdiagnosis of the vegetative state: Retrospective study in a rehabilitation unit. *BMJ, 313*(7048), 13–16.

Anticevic, A., Cole, M. W., Murray, J. D., Corlett, P. R., Wang, X. J., & Krystal, J. H. (2012). The role of default network deactivation in cognition and disease. *Trends in Cognitive Science, 16*(12), 584–592.

Bekinschtein, T., Leiguarda, R., Armony, J., Owen, A., Carpintiero, S., Niklison, J., . . . & Manes, F. (2004). Emotion processing in the minimally conscious state. *Journal of Neurology, Neurosurgery, and Psychiatry, 75*, 788.

Bekinschtein, T., Tiberti, C., Niklison, J., Tamashiro, M., Ron, M., Carpintiero, S., . . . & Manes F. Assessing level of consciousness and cognitive changes from vegetative state to full recovery. *Neuropsychological Rehabilitation, 15*, 307–322.

Boly, M., Faymonville, M. E., Peigneux, P., Lambermont, B., Damas, P., Del Fiore, G., . . . & Laureys, S. (2004). Auditory processing in severely brain injured patients: Differences between the minimally conscious state and the persistent vegetative state. *Archives of Neurology, 61*, 233–238.

Boly, M., Faymonville, M. E., Peigneux, P., Lambermont, B., Damas, F., Luxen, A., . . . & Laureys, S. (2005). Cerebral processing of auditory and noxious stimuli in severely brain injured patients: Differences between VS and MCS. *Neuropsychological Rehabilitation, 15*, 283–289.

Boly, M., Faymonville, M. E., Schnakers, C., Peigneux, P., Lambermont, B., Phillips, C., . . . & Laureys S. (2008). Perception of pain in the minimally conscious state with PET activation: An observational study. *The Lancet Neurology, 7*, 1013–1020.

Boly, M., Garrido, M. I., Gosseries, O., Bruno, M. A., Boveroux, P., Schnakers, C., . . . & Friston, K. (2011). Preserved feedforward but impaired top-down processes in the vegetative state. *Science, 332*(6031), 858–862.

Bouwes, A., Binnekade, J. M., Kuiper, M. A., Bosch, F. H., Zandstra, D. F., Toornvliet, A. C., . . . & Horn, J. (2012). Prognosis of coma after therapeutic hypothermia: A prospective cohort study. *Annals of Neurology, 71*(2), 206–212.

Bruno, M. A., Ledoux, D., Vanhaudenhuyse, A., Gosseries, O., Thibaut, A., & Laureys, S. (2012a). Prognosis of patients with altered state of consciousness. In C. Schnakers & S. Laureys (Eds.). *Coma and disorders of consciousness*. London: Springer-Verlag.

Bruno, M. A., Majerus, S., Boly, M., Vanhaudenhuyse, A., Schnakers, C., Gosseries, O., . . . & Laureys, S. (2012b). Functional neuroanatomy underlying the clinical subcategorization of minimally conscious state patients. *Journal of Neurology, 259*(6), 1087–1098.

Bruno, M. A., Vanhaudenhuyse, A., Thibaut, A., Moonen, G., & Laureys, S. (2011). From unresponsive wakefulness to minimally conscious PLUS and functional locked-in syndromes: Recent advances in our understanding of disorders of consciousness. *Journal of Neurology, 258*(7), 1373–1384.

Candelieri, A., Cortese, M. D., Dolce, G., Riganello, F., & Sannita, W. G. (2011). Visual pursuit: Within-day variability in the severe disorder of consciousness. *Journal of Neurotrauma, 28*(10), 2013–2017.

Casali, A. G., Gosseries, O., Rosanova, M., Boly, M., Sarasso, S., Casali, K. R., . . . & Massimini, M. (2013). A theoretically based index of consciousness independent of sensory processing and behavior. *Science Translational Medicine*, *5*(198), 198ra105.

Cauda, F., Micon, B. M., Sacco, K., Duca, S., D'Agata, F., Geminiani, G., & Canavero, S. (2009). Disrupted intrinsic functional connectivity in the vegetative state. *Journal of Neurology, Neurosurgery, and Psychiatry*, *80*(4), 429–431.

Chatelle, C., Majerus, S., Whyte, J., Laureys, S., & Schnakers, C. (2012). A sensitive scale to assess nociceptive pain in patients with disorders of consciousness. *Journal of Neurology, Neurosurgery, and Psychiatry*, *83*(12), 1233–1237.

Childs, N. L., Mercer, W. N., & Childs, H. W. (1993). Accuracy of diagnosis of persistent vegetative state. *Neurology*, *43*(8), 1465–1467.

Cohen, N. H., & Kummer, H. B. (2006). Ethics update: Lessons learned from Terri Schiavo—The importance of healthcare proxies in clinical decision-making. *Current Opinion in Anaesthesiology*, *19*(2), 122–126.

Crone, J. S., Ladurner, G., Höller, Y., Golaszewski, S., Trinka, E., & Kronbichler, M. (2011). Deactivation of the default mode network as a marker of impaired consciousness: An fMRI study. *PLoS One*, *6*(10), e26373.

Cruse, D., Chennu, S., Chatelle, C., Bekinschtein, T. A., Fernández-Espejo, D., Pickard, J. D., . . . & Owen, A. M. (2011). Bedside detection of awareness in the vegetative state: A cohort study. *The Lancet*, *378*(9809), 2088–2094.

Dolce, G., Lucca, L. F., Candelieri, A., Rogano, S., Pignolo, L., & Sannita, W. G. (2011). Visual pursuit in the severe disorder of consciousness. *Journal of Neurotrauma*, *28*(7), 1149–1154.

Faugeras, F., Rohaut, B., Weiss, N., Bekinschtein, T. A., Galanaud, D., Puybasset, L., . . . & Naccache, L. (2011). Probing consciousness with event-related potentials in the vegetative state. *Neurology*, *77*(3), 264–268.

Fernández-Espejo, D., Bekinschtein, T., Monti, M. M., Pickard, J. D., Junque, C., Coleman, M. R., & Owen, A. M. (2011). Diffusion weighted imaging distinguishes the vegetative state from the minimally conscious state. *NeuroImage*, *54*(1), 103–112.

Fernández-Espejo, D., Junque, C., Bernabeu, M., Roig-Rovira, T., Vendrell, P., & Mercader, J. M. (2010). Reductions of thalamic volume and regional shape changes in the vegetative and the minimally conscious states. *Journal of Neurotrauma*, *27*(7). 1187–1193.

Fernández-Espejo, D., Soddu, A., Cruse, D., Palacios, E. M., Junque, C., Vanhaudenhuyse, A., . . . & Owen, A. M. (2012). A role for the default mode network in the bases of disorders of consciousness. *Annals of Neurology*, *72*(3), 335–343.

Fins, J. J., Schiff, N. D., & Foley, K. M. (2007). Late recovery from the minimally conscious state: Ethical and policy implications. *Neurology*, *68*, 304–307.

Giacino, J., Ashwal, S., Childs, N., Cranford, R., Jennett, B., Katz, D., . . . & Zafonte, R. (2002). The minimally conscious state: Definition and diagnostic criteria. *Neurology*, *58*, 349–353.

Giacino, J. T., Kalmar, K., & Whyte, J. (2004). The JFK Coma Recovery Scale-Revised: Measurement characteristics and diagnostic utility. *Archives of Physical Medicine and Rehabilitation*, *85*(12), 2020–2029.

Gill-Thwaites, H. (2006). Lotteries, loopholes and luck: Misdiagnosis in the vegetative state patient. *Brain Injury*, *20*(13-14), 1321–1328.

Gosseries, O., Demertzi, A., Ledoux, D., Bruno, M. A., Vanhaudenhuyse, A., Thibaut, A., . . . & Schnakers, C. (2012). Burnout in healthcare workers managing chronic patients with disorders of consciousness. *Brain Injury, 26*(12), 1493–1499.

Jennett, B. (2005). The assessment and rehabilitation of vegetative and minimally conscious patients: Definitions, diagnosis, prevalence and ethics. *Neuropsychological Rehabilitation, 15*. 163–165.

Jennett, B., & Plum, F. (1972). Persistent vegetative state after brain damage: A syndrome in search of a name. *Lancet, 1*, 734–737.

Landsness, E., Bruno, M. A., Noirhomme, Q., Riedner, B., Gosseries, O., Schnakers, C., . . . & Boly, M. (2011). Electrophysiological correlates of behavioural changes in vigilance in vegetative state and minimally conscious state. *Brain, 134*(8), 2222–2232.

Laureys, S. (2005). Science and society: Death, unconsciousness and the brain. *Nature Reviews Neuroscience, 6*(11), 899–909.

Laureys, S., Celesia, G. G., Cohadon, F., Lavrijsen, J., León-Carrión, J., Sannita, W. G., . . . & Dolce, G. (2010). Unresponsive wakefulness syndrome: A new name for the vegetative state or apallic syndrome. *BMC Medicine, 8*, 68.

Laureys, S., Faymonville, M. E., Degueldre, C., Fiore, G. D., Damas, P., Lambermont, B., . . . & Maquet, P. (2000a). Auditory processing in the vegetative state. *Brain, 123*, 1589–1601.

Laureys, S., Faymonville, M. E., Luxen, A., Lamy, M., Franck, G., & Maquet, P. (2000b). Restoration of thalamocortical connectivity after recovery from persistent vegetative state. *The Lancet, 355*, 1790–1791.

Laureys, S., Faymonville, M. E., Peigneux, P., Damas, P., Lambermont, B., Del Fiore, G., . . . & Franck, G. (2002). Cortical processing of noxious somatosensory stimuli in the persistent vegetative state. *NeuroImage, 17*, 732–741.

Laureys, S., Goldman, S., Phillips, C., Van Bogaert, P., Aerts, J., Luxen, A., . . . & Maquet, P. (1999a). Impaired effective cortical connectivity in vegetative state: Preliminary investigation using PET. *NeuroImage, 9*, 377–382.

Laureys, S., Lemaire, C., Maquet, P., Phillips, C., & Franck, G. (1999b). Cerebral metabolism during vegetative state and after recovery to consciousness. *Journal of Neurology, Neurosurgery, and Psychiatry, 67*, 121.

Laureys, S., Perrin, F., Schnakers, C., Boly, M., & Majerus, S. (2005). Residual cognitive function in comatose, vegetative and minimally conscious states. *Current Opinion in Neurology, 18*, 726–733.

Leonardi, M., Giovannetti, A. M., Pagani, M., Raggi, A., & Sattin, D. (2012). Burden and needs of 487 caregivers of patients in vegetative state and in minimally conscious state: Results from a national study. *Brain Injury, 26*(10), 1201–1210.

Luchetti, M. (2010). Eluana Englaro, chronicle of a death foretold: Ethical considerations on the recent right-to-die case in Italy. *Journal of Medical Ethics, 36*(6), 333–335.

Lull, N., Noé, E., Lull, J. J., García-Panach, J., Chirivella, J., . . . & Robles, M. (2010). Voxel-based statistical analysis of thalamic glucose metabolism in traumatic brain injury: Relationship with consciousness and cognition. *Brain Injury, 24*(9), 1098–1107.

McNett, M. (2007). A review of the predictive ability of Glasgow Coma Scale scores in head-injured patients. *Journal of Neuroscience Nursing, 39*(2), 68–75.

Monti, M. M., Vanhaudenhuyse, A., Coleman, M. R., Boly, M., Pickard, J. D., Tshibanda, L., . . . & Laureys, S. (2010). Willful modulation of brain activity in disorders of consciousness. *New England Journal of Medicine, 362*(7), 579–589.

The Multi-Society Task Force on Persistent Vegetative State. (1994). Medical aspects of the persistent vegetative state. *New England Journal of Medicine, 330*(21), 1499–1508.

Newcombe, V. F., Williams, G. B., Scoffings, D., Cross, J., Carpenter, T. A., Pickard, J. D., & Menon, D. K. (2010). Aetiological differences in neuroanatomy of the vegetative state: Insights from diffusion tensor imaging and functional implications. *Journal of Neurology, Neurosurgery, and Psychiatry, 81*(5), 552–561.

Owen, A. M., & Coleman, M. R. (2008). Detecting awareness in the vegetative state. *Annals of the New York Academy of Sciences, 1129*, 130–138.

Phillips, C. L., Bruno, M. A., Maquet, P., Boly, M., Noirhomme, Q., Schnakers, C., . . . & Laureys, S. (2011). "Relevance vector machine" consciousness classifier applied to cerebral metabolism of vegetative and locked-in patients. *NeuroImage, 56*(2), 797–808.

Posner, J., Saper, C., Schiff, N., & Plum, F. (2007). *Plum and Posner's diagnosis of stupor and coma*. New York: Oxford University Press.

Rosanova, M., Gosseries, O., Casarotto, S., Boly, M., Casali, A. G., Bruno, M. A., . . . & Massimini, M. (2012). Recovery of cortical effective connectivity and recovery of consciousness in vegetative patients. *Brain, 135*(Pt 4), 1308–1320.

Schiff, N. D., Rodriguez-Moreno, D., Kamal, A., Kim, K. H., Giacino, J. T., Plum, F., & Hirsch, J. (2005). fMRI reveals large-scale network activation in minimally conscious patients. *Neurology, 64*, 514–523.

Schnakers, C., Chatelle, C., Demertzi, A., Majerus, S., & Laureys, S. (2012). What about pain in disorders of consciousness? *AAPS Journal, 14*(3), 437–444.

Schnakers, C., Chatelle, C., Vanhaudenhuyse, A., Majerus, S., Ledoux, D., Boly, M., . . . & Laureys, S. (2010). The Nociception Coma Scale: A new tool to assess nociception in disorders of consciousness. *Pain, 148*(2), 215–219.

Schnakers, C., Giacino, J., Kalmar, K., Piret, S., Lopez, E., Boly, M., . . . & Laureys, S. (2006). Does the FOUR score correctly diagnose the vegetative and minimally conscious states? *Annals of Neurology, 60*(6), 744–745.

Schnakers, C., Laureys, S., & Majerus, S. (2011). I am conscious: Mr ASL says so. *Neurology, 77*(16), 1506–1507.

Schnakers, C., Perrin, F., Schabus, M., Majerus, S., Ledoux, D., Damas, P., . . . & Laureys, S. (2008). Voluntary brain processing in disorders of consciousness. *Neurology, 71*(20), 1614–1620.

Schnakers, C., Vanhaudenhuyse, A., Giacino, J., Ventura, M., Boly, M., Majerus, S., . . . & Laureys, S. (2009). Diagnostic accuracy of the vegetative and minimally conscious state: Clinical consensus versus standardized neurobehavioral assessment. *BMC Neurology, 9*, 35.

Seel, R. T., Sherer, M., Whyte, J., Katz, D. I., Giacino, J. T., Rosenbaum, A. M., . . . & Zasler, N. (2010). Assessment scales for disorders of consciousness: Evidence-based recommendations for clinical practice and research. *Archives of Physical Medicine and Rehabilitation, 91*(12), 1795–1813.

Silva, S., Alacoque, X., Fourcade, O., Samii, K., Marque, P., Woods, R., . . . & Loubinoux, I. (2010). Wakefulness and loss of awareness: Brain and brainstem interaction in the vegetative state. *Neurology, 74*, 313–320.

Teasdale, G., & Jennett, B. (1976). Assessment and prognosis of coma after head injury. *Acta Neurochirurgica (Wien), 34*(1-4), 45–55.

Tommasino, C., Grana, C., Lucignani, G., Torri, G., & Fazio, F. (1995). Regional cerebral metabolism of glucose in comatose and vegetative state patients. *Journal of Neurosurgical Anesthesiology, 7*, 109–116.

Vanhaudenhuyse, A., Noirhomme, Q., Tshibanda, L. J., Bruno, M. A., Boveroux, P., Schnakers, C., . . . & Boly, M. (2010). Default network connectivity reflects the level of consciousness in non-communicative brain-damaged patients. *Brain, 133*, 161–171.

Ward, L. M. (2011). The thalamic dynamic core theory of conscious experience. *Consciousness and Cognition, 20*(2), 464–486.

7

Decoding Thoughts in Behaviorally Nonresponsive Patients

ADRIAN M. OWEN AND LORINA NACI

INTRODUCTION

In recent years, improvements in intensive care have led to an increase in the number of patients who survive serious brain injury. Some go on to make a good recovery, but many do not, and following a period of coma, may progress to either a vegetative or minimally conscious state. A diagnosis of vegetative state is made after repeated examinations have yielded no evidence of sustained, reproducible, purposeful or voluntary behavioral response to visual, auditory, tactile or noxious stimuli. After a traumatic brain injury, the vegetative state is considered permanent, and thus irreversible, when it lasts longer than 1 year. For non-traumatic cases, for which the prognosis is generally worse, the duration is 6 months (Royal College of Physicians, 2003), or 3 months according to US guidelines (The Multi-Society Task Force on PVS, 1994). The minimally conscious state is a relatively new diagnostic category (Giacino et al., 2002), and describes patients who show inconsistent, but reproducible evidence of awareness. In both cases, assessment is extremely difficult and the diagnosis relies on subjective interpretation of observed behavior. These difficulties, coupled with inadequate experience and knowledge engendered through the relative rarity of these complex conditions, contribute to an alarmingly high rate of misdiagnosis (up to 43%) in these patient groups (Childs et al., 1993; Andrews et al., 1996; Schnakers et al., 2009).

Recent advances in neuroimaging technology have suggested a number of possible solutions to this problem. In this chapter, recent studies that have used both functional magnetic resonance imaging (fMRI) and electroencephalography (EEG) in this context will be reviewed. The results suggest an urgent need for a re-evaluation of the existing diagnostic guidelines for behaviorally non-responsive patients (including the vegetative state and related disorders of consciousness) and for the development and formal inclusion of validated, standardized neuroimaging procedures into those guidelines.

USING fMRI TO DECODE THOUGHTS

In recent years, many attempts have been made to "decode" mental decisions or thoughts in healthy participants (e.g. Weiskopf et al., 2004; Boly et al., 2007; Haynes et al., 2007; Cerf et al., 2010; Gallivan et al., 2011). The principle often employed capitalizes on the fact that certain types of thought are associated with a unique brain activation pattern that can be used as a signature for that specific thought. If a statistical classifier is trained to recognize these characteristic signatures, a volunteer's thoughts can be decoded (within the constraints of the experimental design) using their brain activity alone.

To achieve acceptable levels of accuracy these methods often rely on mental imagery as a proxy for the physical response being decoded. For example, in one early study, four non-naive participants learned, with the aid of feedback, to willfully regulate their fMRI signal using self-chosen visual imagery strategies (e.g. pictures of buildings, spatial navigation, clenching, dancing) (Weiskopf et al., 2004). In a more sophisticated design, information derived from both the timing (onset and offset) and the source location of the hemodynamic response was used to decode which of four possible answers was being given to questions (Sorger et al., 2009). To indicate their choice (or "thought"), participants imagined one of two tasks, beginning at one of four times and continuing for different pre-specified durations. An automated decoding procedure deciphered the answer by analyzing the single-trial BOLD responses in real time with a mean accuracy of 94.9%.

In one large study (Boly et al., 2007), 34 healthy volunteers were asked to imagine hitting a tennis ball back and forth to an imaginary coach when they heard the word "tennis" (thereby eliciting vigorous imaginary arm movements) and to imagine walking from room to room in their house when they heard the word "house" (thereby eliciting imaginary spatial navigation). Imagining playing tennis was associated with robust activity in the supplementary motor area in each and every one of the participants scanned. In contrast, imagining moving from room to room in a house activated the parahippocampal cortices, the posterior parietal lobe and the lateral premotor cortices; all regions that have been shown to contribute to imaginary, or real, spatial navigation (Aguirre et al., 1996). By simply examining the responses elicited during the imagery tasks, Boly and colleagues (2007) were able to "decode" which task was being mentally "performed." Moreover, the robustness and reliability of fMRI responses across individuals meant that activity in these regions could be used to confirm that the participants retained the ability to understand instructions, to carry out different mental tasks in response to those instructions and, therefore, were able to exhibit willed, voluntary brain behavior in the absence of any overt action. On this basis, Boly and colleagues (2007) argued that, like any other form of action that requires a choice between one of several possible responses, these brain responses require *awareness*; that is to say, awareness of the various contingencies that govern the relationship between a given stimulus (in this case, the cue word for one of two possible imagery tasks) and a response (in this case, imagining a type of action). Put simply, fMRI responses of this sort can be used to measure awareness because awareness is necessary for them to occur. Moreover, aside from the presence of awareness, these type of fMRI responses can be used to demonstrate a participant's free choice and executive control that is needed to orchestrate the requisite mental functions (e.g. language comprehension, long-term and working memory, attention) for successful task performance. Choice and control are two aspects of *self-motivated activity*, a trait central to "personhood," or to the status of being a person who can exhibit willful behavior (Warren, 1973).

Owen et al. (2006; 2007) used this same logic to demonstrate that a young woman who fulfilled all internationally agreed criteria for the vegetative state was, in fact, consciously aware and able to make responses of this sort using her brain activity. The patient, who was involved in a complex road traffic accident and had sustained very severe traumatic brain injuries, had remained entirely unresponsive for a period of 6 months prior to the fMRI scan. During two different scanning sessions, the patient was instructed to perform the two mental imagery tasks described above. In each case she was asked to imagine playing tennis/moving around the rooms of her home (for 30 seconds) when she heard the word *tennis/house*, and to relax (for 30 seconds) when she heard the word *relax*. When she was asked to imagine playing tennis (Figure 7.1, **patient 5**), significant activity was observed repeatedly in the supplementary motor area (Owen et al., 2006) that was indistinguishable from that observed in the healthy volunteers scanned by Boly et al. (2007). Moreover, when she was asked to imagine walking through her home, significant activity was observed in the parahippocampal gyrus, the posterior parietal cortex and the lateral premotor cortex which was again, indistinguishable from those observed in healthy volunteers (Owen et al., 2006; 2007). The patient's brain activity was statistically robust, reproducible, task appropriate (enhanced following the "tennis"/"-house" cue and returning to baseline following the "relax" cue), sustained over long time intervals (30 seconds), and repeated over each 5-minute session. On

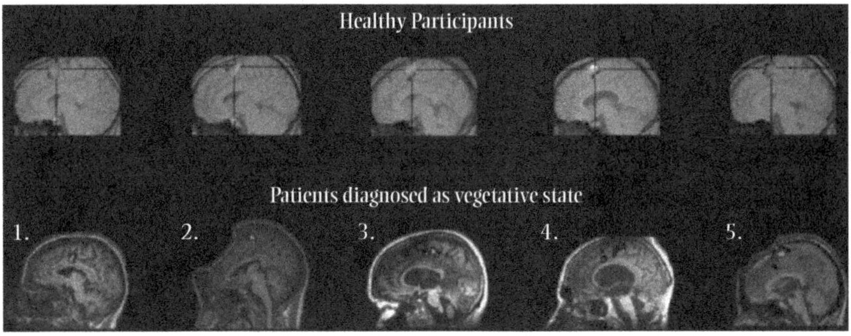

Figure 7.1 Top row: Five healthy participants asked to imagine playing tennis in the fMRI scanner (adapted from Boly et al., 2007). In all five cases, significant activity was observed in the premotor cortex, indicating that they had understood the instruction and were responding by carrying out the appropriate type of mental imagery; that is, *following a command.* Bottom row: Formally identical responses in five patients who behaviorally meet the clinical criteria for a diagnosis of vegetative state (adapted from Owen et al. (2006) {Patient 5} and Monti/Vanhaudenhuyse et al. (2010) {Patients 1-4}) confirming that, in spite of an inability to respond physically, these patients can still demonstrate *command following* by modulating their cortical fMRI activity. Such responses are observed in approximately 17% of vegetative patients.

this basis, it was concluded that, despite fulfilling all of the clinical criteria for a diagnosis of vegetative state, this patient retained the ability to understand spoken commands and to respond to them through her brain activity, rather than through speech or movement, confirming that she was consciously aware of herself and her surroundings. In a follow-up study of 23 patients who were behaviorally diagnosed as vegetative, Monti/Vanhaudenhuyse et al. (2010) showed that 4 (17%) were able to generate reliable responses of this sort in the fMRI scanner (Figure 7.1, **patients 1-4**).

After a severe brain injury, when the request to move a hand or a finger is followed by an appropriate motor response, the diagnosis can change from vegetative state (no evidence of awareness) to minimally conscious state (some evidence of awareness). By analogy then, we have argued that if the request to activate, say, the supplementary motor area of the brain by imagining moving the hand is followed by an appropriate brain response, we should give that response the very same weight (Owen and Coleman, 2008a; Owen, 2013; Fernández-Espejo and Owen, 2013). Skeptics may argue that brain responses are somehow less physical, reliable or immediate than motor responses but, as is the case with motor responses, all of these arguments can be dispelled with careful measurement, replication and objective verification (Owen et al., 2006; Boly et al., 2007; Owen et al., 2007; Monti/Vanhaudenhuyse et al., 2010;

Naci et al., 2013; Naci and Owen, 2013; Fernández-Espejo and Owen, 2013; Hampshire et al., 2013). For example, if a patient who was assumed to be unaware raised his/her hand to command on just one occasion, there would remain some doubt about the presence of awareness given the possibility that this movement was a chance occurrence, coincident with the instruction. However, if that same patient were able to repeat this response to command on ten occasions, there would remain little doubt that the patient was aware. By the same token, if that patient was able to activate his/her supplementary motor area in response to command (e.g. by being told to imagine hand movements), and was able to do this on every one of ten trials, we would have to accept that this patient was consciously aware. Like most neuroimaging investigations, replication of this sort was inherent in both of the studies described above (Owen et al., 2006; Monti/Vanhaudenhuyse et al., 2010), because the statistically significant results depended on multiple, similar responses being exhibited across repeated trials.

Owen and Coleman (2008b) extended the general principles discussed above, by which active mental rehearsal is used to signify awareness, to show that communication of "yes" and "no" responses was possible using the same approach. Thus, a healthy volunteer was able to reliably convey a "yes" response by imagining playing tennis and a "no" response by imaging moving around a house, thereby providing the answers to simple questions posed by the experimenters using only their brain activity. This technique was further refined by Monti/Vanhaudenhuyse et al. (2010) who successfully decoded three "yes" and "no" responses from each of 16 healthy participants with 100% accuracy using only their real time changes in the supplementary motor area (during tennis imagery) and the parahippocampal place area (during spatial navigation). Moreover, in one traumatic brain injury patient, who had been repeatedly diagnosed as vegetative over a five-year period, similar questions were posed and successfully decoded using the same approach (Monti/Vanhaudenhuyse et al., 2010). Thus, this patient was able to convey biographical information that was not known to the experimenters at the time (but was verified as factually correct) such as his father's name and the last place that he had visited on vacation before his accident 5 years earlier. In contrast, and despite a re-classification to minimally conscious state following the fMRI scan, it remained impossible to establish any form of communication with this patient at the bedside.

An obvious application for approaches of this sort is to begin to involve some of these patients in the decision-making processes involved in their own therapeutic care and management. To date, this has only been achieved successfully in one patient, who had been repeatedly diagnosed as vegetative for twelve years following a traumatic brain injury (Fernández-Espejo and Owen, 2013). The patient was a male who, in December 1999 and at the age of 26, had suffered a severe closed head injury in a motor-vehicle accident. On admission to hospital he had a Glasgow Coma Scale (Teasdale and Jennett, 1974) score of 4, meaning that he was unable to open his eyes or produce any sound, and his only response was extension to painful stimulation. Over the next twelve years, the patient was assessed regularly by experienced Neurologists and multidisciplinary teams and throughout this period his behavior remained consistent with the internationally accepted criteria for the vegetative state. Indeed, over one 14-month period in 2011-2013, a total of 20 standardized behavioral assessments were performed by a multidisciplinary team, at different times of the day and in different postural positions, using the Coma Recovery Scale—Revised (Giacino et al., 2004), and his diagnosis was unchanged throughout. In February 2012, twelve years and two months after his accident, the patient was first scanned using the fMRI mental imagery approach described above (Owen et al., 2006; Monti/Vanhaudenhuyse et al., 2010). The patient was able to provide correct answers to multiple externally verifiable questions, including his own name, his whereabouts, the name of his personal support worker (who he had only encountered in the years following his accident), the current date, and other basic factual information (e.g. whether a banana is yellow). Two non-verifiable questions were then posed, including one pertaining to his care preferences (e.g. whether he liked watching (ice) hockey games on TV), and another to details about his current clinical condition (e.g. whether he was in any physical pain). Within the time-constrains of the scanning visits, the majority of responses to these questions were verified in independent sessions that posed the reverse questions (e.g. "Is your name Mike?" vs. "Is your name Scott?"). To date, answers to 12 different questions have been obtained across several sessions, despite the fact that the patient remains entirely physically non-responsive at the bedside (Fernández-Espejo and Owen, 2013).

Although techniques like the ones described above, require that the patient engages in rather specific types of mental imagery (playing tennis or moving from room to room through a house), that is not really the main point that allows consciousness to be detected and communication to occur. All that is required to detect consciousness is a reliable indicator that a patient can turn his or her attention to a specific scenario, because this then serves as a "neural proxy" for a physical "response to command." By extension, if it can be shown that the patient can turn his or her attention to two separate scenarios, then communication is possible because those two separate scenarios can be linked to "yes" responses and "no" responses, respectively. Thus, mental imagery is not necessary at all, but serves as a simple vehicle for guiding a patient's attention one way or another.

A related and possibly simpler approach to detecting covert awareness after brain injury, therefore, is to target processes that require the wilful adoption of "mind-sets" in carefully matched (perceptually identical) experimental and control conditions. For example, Monti et al. (2009), presented healthy volunteers with a series of neutral words, and alternatively instructed them to just listen, or to count, the number of times a given word was repeated. As predicted, the counting task revealed the fronto-parietal network that has been previously associated with target detection and working memory. When tested on this same procedure, a severely brain injured patient produced a very similar pattern of activity, confirming that he could willfully adopt differential mind-sets as a function of the task conditions and could actively maintain these mind-sets across time. These covert abilities were entirely absent from his documented behavioral repertoire. As in the tennis/spatial navigation examples described above, because the external stimuli (a series of words) were identical in the two conditions any difference in brain activity observed cannot reflect an "automatic" brain response (i.e. one that can occur in the absence of consciousness). Rather, the activity must reflect the fact that the patient has performed a particular action (albeit a "brain action") in response to the stimuli on one (but not the other) presentation; in this sense, the brain response is entirely analogous to a (motor) response to command and should carry the same weight as evidence of awareness.

Naci and colleagues (Naci et al., 2013; Naci and Owen, 2013) took this general principle even further and developed a novel tool for communicating with non-responsive patients based on how they selectively directed their attention to sounds while in the fMRI scanner. It is well established that selective attention can significantly enhance the neural representation of attended sounds (Bidet-Caulet et al., 2007), although most previous studies have focused on group-level changes rather than individual responses that are crucial for work with (individual) brain-injured patients. In their first study (Naci et al., 2013), fifteen healthy volunteers answered questions (e.g., "Do you have brothers or sisters?") in the fMRI scanner, by selectively attending to the appropriate word ("yes" or "no"), which was played to them auditorily, interspersed with "distractor" stimuli (digits 1-9). Ninety percent of the answers were decoded correctly based on activity changes within the attention network of the brain. Moreover, the majority of volunteers conveyed their answers with less than 3 min of scanning, which represents a significant time saving over the mental imagery methods described above (Owen et al., 2006; 2007; Boly et al., 2007). Indeed, a formal comparison between the two approaches revealed improved individual success rates and an overall reduction in the scanning times required to correctly detect responses; 100% of volunteers showed significant task-appropriate activity to the selective attention task, compared to 87% to the motor imagery. This result is consistent with previous studies showing that a proportion of healthy volunteers do not produce reliable brain activation during mental imagery tasks (Boly et al., 2007).

In a follow up study, Naci and Owen (2013) used the same approach to test for residual conscious awareness and communication abilities in three behaviorally nonresponsive, brain-injured patients. As in the previous study of healthy participants, the patients had to either "count" or "relax" as they heard a sequence of sounds. The word *count* at the beginning of the sequence instructed the patient to count the occurrences of a target word (*yes* or *no*), while the word *relax* instructed them to relax and ignore the sequence of words. Reliable activity increases in the attention network of the brain after the word *count* relative to the word *relax* was taken as evidence of command following. All three patients (two of whom were diagnosed as being in a minimally conscious state and one as being in a vegetative state) were able to convey their ability to follow commands inside the fMRI scanner by following the instructions in this way. In stark contrast, extremely limited or a complete lack of behavioral responsivity was observed in repeated bedside assessments of all three patients. These results confirm that selective attention is an appropriate vehicle for detecting covert awareness in some behaviorally non-responsive patients who are presumed to mostly or entirely lack any cognitive abilities whatsoever.

In a following series of scans, communication was attempted in two of the patients. The communication scans were similar to those in the command-following scan, with one exception. Instead of an instruction (count or relax), a binary question (e.g. "Is your name Steven?") preceded each sound sequence. Thus, each patient then had to willfully choose which word to attend to (count) and which to ignore, depending on which answer he wished to convey to the specific question that had been asked. Using this method, the two patients (one diagnosed as minimally conscious state and one diagnosed as vegetative state) were able to use selective attention to repeatedly communicate correct answers to questions that were posed to them by the experimenters (Naci and Owen, 2013). In the absence of external cues as to which word the patient was attending to, the functional brain activation served as the only indicator of the patient's intentions—and in both cases, led to the correct answers being decoded. For example, when asked, "Are you in a supermarket?" one patient showed significantly more activation for "no" than "yes" sequences in a network of brain areas that had been previously activated when that patient was focusing attention on external cues (Figure 7.2). Conversely, when asked, "Are you in a hospital?" the patient showed significantly more activation for "yes" than "no" sequences in the same regions. Despite his diagnosis

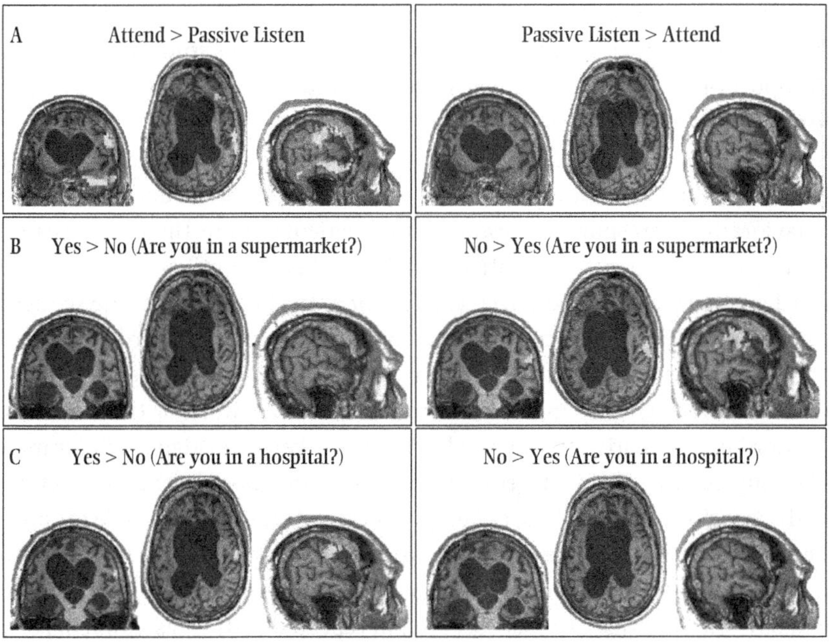

Figure 7.2 Command-following (A) and communication (B and C) scans in patient 3, clinically diagnosed as being in a vegetative state. Brain activity is overlaid on the patient's native anatomic volume. The opposite directions of each contrast (i.e., $a > b$ or $b > a$) are shown on the left and right sides of each panel. A) The command-following scan also served to localize the brain foci of attention unique to the patient. B and C) Selective attention to the answer word (either yes or no) during each communication scan was investigated within these regions. Attention to the answer in each question (B, no; C, yes) significantly activated the precentral or motor region.

(vegetative state for 12 years), the fMRI approach allowed this patient to establish interactive communication with the research team in 4 different fMRI sessions. The patient's brain responses within specific regions were remarkably consistent and reliable across 2 different scanning visits, 5 months apart, during which the patient maintained the long-standing vegetative state diagnosis. For all 4 questions, the patient produced a robust neural response and was able to provide the correct answer with 100% accuracy. The patient's brain activity in the communication scans not only further corroborated that he was, indeed, consciously aware but also revealed that he had far richer cognitive reserves than could be assumed based on his clinical diagnosis. In particular, beyond the ability to pay attention, these included autobiographical knowledge and awareness of his location in time and space.

These types of approaches all illustrate a paradigmatic shift towards the use of active (e.g. willful) tasks in the assessment of covert awareness after serious brain injury. What sets such tasks apart is that the neural responses required are not produced *automatically* by the eliciting stimulus, but rather, depend on time-dependent and sustained responses generated by the participants themselves. Such behavior (albeit neural "behavior") provides a proxy for a motor action and is, therefore, an appropriate vehicle for reportable awareness (Zeman, 2009).

USING EEG TO DECODE THOUGHTS

Performing fMRI in severely brain-injured patients is enormously challenging; in addition to considerations of cost and scanner availability, the physical stress incurred by patients as they are transferred to a suitably equipped fMRI facility is significant. Movement artifacts often occur in imaging datasets from patients who are unable to remain still, while metal implants, including plates and pins which are common in many traumatically-injured populations, may rule out fMRI altogether. EEG measures the activity of groups of cortical neurons from scalp electrodes and is far less expensive than fMRI, both in terms of initial cost and maintenance. EEG recordings are unaffected by any resident metallic implants and, perhaps most importantly, can be used at the bedside (Vaughan et al., 2006). In brain-injured patients, EEG recordings are typically made in the acute period and allow for broad assessments of cortical damage including the occurrence of brain death. However, uncertainty about the causes of abnormal raw EEG patterns (i.e. damage to the cortex itself, or to sub-cortical structures which influence cortical activity) provides challenges for its use as a more precise tool for the assessment of awareness (Kulkarni et al., 2007).

Motor imagery produces clearly distinguishable modulation of EEG sensorimotor rhythms (SMR) (Wolpaw et al., 1991; Cincotti et al., 2003) similar to those seen during motor execution, and has been the basis of several recent attempts to detect conscious awareness after severe brain injury. For example, in one early study, Kotchoubey and colleagues (2003) described a completely locked-in patient whose slow EEG activity differed significantly between trials when he was asked to "try" to move the left, as compared to the right, hand. In the EEG record, imagined movements (motor imagery) are evident in the form of reductions of power—or event-related desynchronisations (ERD)—of the mu (~7-13Hz) and/or beta (~13-30Hz) bands over the topographically appropriate regions of the motor cortex—for example, over the lateral premotor cortex for hand movements and over more medial premotor cortex for toe movements (Pfurtscheller and Neuper, 1997). In some individuals, these ERDs may also be accompanied by event-related synchronizations (ERS; relative increases in power) over motor areas contralateral to, or surrounding, the ERD (Pfurtscheller et al., 2006; 2008). Using classification techniques it is now possible, on the basis of these EEG responses alone, to determine the form of motor imagery being performed by a conscious individual with a high degree of accuracy (Guger et al., 2003). For example, Cruse et al. (2011) recently reported a new EEG-based classification technique in which two mental imagery responses (squeezing the right hand or squeezing the toes) were successfully decoded offline in 9 out of 12 healthy individuals with accuracy rates varying between 60-91%. The same approach was then used to attempt to detect evidence of command following in the absence of any overt behavior, in a group of 16 patients who met the internationally agreed criteria for a diagnosis of vegetative state. Three of these patients (19%; 2 traumatic brain injury and 1 non-traumatic brain injury) were repeatedly and reliably able to generate appropriate EEG responses to the two distinct commands ("squeeze your right hand" or "squeeze your toes"), despite being behaviorally entirely unresponsive, indicating that they were aware and following the task instructions. Indeed, on the basis of such data, far broader conclusions about residual cognition can be drawn. For example, performance of this complex task makes multiple demands on many cognitive functions, including sustained attention (over 90-second blocks), response selection (between the two imagery tasks), language comprehension (of the task instructions) and working memory (to remember which task to perform across multiple trials within each block)—all aspects of "top-down" cognitive control that are usually associated with—indeed, could be said to characterize—normal conscious awareness (Naccache, 2006).

In a follow-up study (Cruse et al., 2012a), twenty-three minimally conscious state patients (15 traumatic brain injury and 8 non-traumatic brain injury) completed the same motor-imagery EEG task. Consistent and robust responses to command were observed in the EEG of 22% of the minimally conscious state patients (5/23). Etiology had a significant impact on the ability to successfully complete this task, with 33% of traumatic patients (5/15) returning positive EEG outcomes, compared with none of the non-traumatic patients (0/8). However, the link between etiology and projected neuroimaging outcomes remains poorly understood and must be interpreted with caution where individual patients are concerned, as patients in both traumatic and non-traumatic groups vary widely in etiologies, neuropathological, and clinical features. Indeed, in some cases non-traumatic brain injured patients return positive outcomes, such as one of the three patients in the aforementioned study by Cruse et al., (2011).

In a more recent study, Cruse et al. (2012b) refined their EEG approach using a simpler and more clinically viable paradigm that required participants to actually try to move their hands and, unlike the two previous studies (Cruse et al., 2011; 2012a), 100% of the healthy volunteers showed reliable ERD and ERS responses (Cruse et al., 2012b). Moreover, in one of the patients studied by Naci and Owen (2013), who had been repeatedly diagnosed as vegetative for 12 years, reliable modulations of sensorimotor beta rhythms were observed following commands to try to move and these could be classified significantly at a single-trial level (see Figure 7.3). This patient is the first published case of a clinically vegetative patient in whom awareness has been demonstrated using two independent imaging methods (fMRI and EEG) in the absence of any supportive evidence from clinical (behavioral) examination (for a complete description of this case, see Fernández-Espejo and Owen, 2013).

Is it possible that appropriate patterns of activity could be elicited in patients like this in the absence of awareness? Could they somehow reflect an "automatic" response to aspects of the task instructions, such as the words "right-hand" and "toes," and not a conscious and overt "action" on the part of

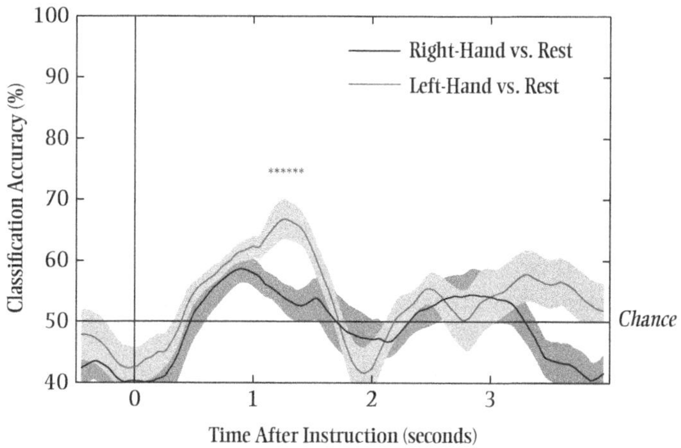

Figure 7.3 EEG activity demonstrating command-following and awareness in a patient diagnosed as being in a vegetative state for 12 years. The plot shows the time-courses of classification accuracies (versus rest) for the trials when the patient, described in detail in this review and diagnosed as vegetative for 12 years, was ask to move his right-hand and left-hand. Lines show means of the 10-fold smoothed classification accuracies. Shaded areas show ±1 standard errors. Stars denote time-points with significantly above chance classification for left-hand vs. rest ($p<.025$). When the frequency band employed in the single-trial classification procedure was narrowed to only that which produced a significant event-related synchronization for left-hand trials (high-beta: 25–30 Hz), significantly above chance classifiability was established for right-hand, as well as left-hand, trials. These data confirmed that this patient was, in fact, aware and able to follow task instructions to (attempt to) move his left and right hands, despite there being no detectable physical response to command. For detailed methods, see Cruse et al., 2012b).

the patient? This is extremely unlikely for a number of reasons. First, the task instructions were delivered once at the beginning of each block of tones that signaled the time to begin each imagery trial. Any "automatic" response to the previously presented verbal instruction would then have to abate and recur in synchrony with these tones/cues that carried no information in themselves about the task to be performed. Indeed, 75% of the healthy control participants tested in the study by Cruse et al. (2011), returned positive EEG outcomes when completing this motor imagery task. However, when these same individuals were instructed *not* to follow the commands—i.e. not to engage in motor imagery—not one participant returned a positive EEG outcome. Evidently, any automatic brain responses generated by listening to the instructions are not sufficient for significant task performance; rather, an act of consistently timed, volitional command following is required. In this context then, it is clear that successful performance of these EEG tasks represents a significant cognitive feat, not only for those patients who were presumed to be vegetative, but also for healthy control participants. That is to say, to be deemed successful, each respondent must have consistently generated the requested mental states to command for a prolonged period of time within each trial, and must have consistently done so across numerous trials. Indeed, one behaviorally vegetative patient was able to produce EEG-responses that were classified with a success rate of 78% (Cruse et al., 2011). In other words, consistently appropriate EEG responses were generated across approximately 100 trials. Conversely, when assessed behaviorally using accepted, standard clinical measures that were administered by experienced, specialist teams, none of these patients exhibited any signs of awareness, including visual fixation, visual pursuit or localization to pain. These results demonstrate that consistent responses to command—a reliable and universally accepted indicator that a patient is not vegetative—need not be expressed behaviorally at all, but rather, can be determined accurately on the basis of EEG responses (Cruse et al., 2012b).

The success of recent EEG techniques for detecting awareness in non-responsive patients (Cruse et al., 2011; 2012a; 2012b), paves the way for the development of a true "brain-computer interface" (BCI) (Birbaumer, 2006)—or simple, reliable communication devices—in this patient group. It seems likely that such devices will provide a form of external control and communication based on mappings of distinct mental states—for example, attempting right-hand movements to communicate "yes," and toe movements to communicate "no" (Cruse et al., 2012b). Indeed, the degrees of freedom provided by EEG have the potential to take this beyond the sorts of binary responses that have worked well using fMRI (Monti/Vanhaudenhuyse et al., 2010; Naci et al., 2013; Naci and Owen, 2013; Fernández-Espejo and Owen, 2013), to allow methods of communication that are far more functionally expressive, based on multiple forms of mental state classification (Farwell and Donchin, 1988; Wolpaw et al., 2002; Sellers and Donchin, 2006). The development of techniques for the real-time classification of these forms of mental imagery (e.g. Cruse et al., 2011; 2012a; 2012b), will open the door for routine two-way communication with some of these patients, ultimately allowing them (within the constrains of BCI technologies) to share information about their inner worlds, experiences and needs.

IMPLICATIONS

Diagnostic Implications

An obvious clinical consequence of the emergence of novel neuroimaging techniques that permit the identification of covert awareness and communication in the absence of any behavioral response is the possibility of improved diagnosis after severe brain injury. It is notable that in one of the cases described above (Cruse et al., 2012b; Naci and Owen, 2013; Fernández-Espejo and Owen, 2013), the patient was repeatedly and rigorously assessed by experienced teams and showed no behavioral sign of awareness on any of these occasions—indeed, this continued to be the case even after awareness had been established unequivocally with both fMRI and EEG. Technically however, he was not *misdiagnosed* (as vegetative), in the sense that any error of judgment was made, because the accepted diagnostic criteria are based on behavior, and no behavioral marker of awareness was missed. Nevertheless, the existing criteria did not accurately capture his actual state of awareness and, in this sense, his vegetative state diagnosis was clearly incorrect. What then, is the appropriate diagnostic label for such patients, who can follow commands with a measurable brain response, but physically remain entirely non-responsive? The term "non-behavioral minimally conscious state" has been suggested (Fins and Schiff, 2006), although because attention, language comprehension and working memory are demonstrably preserved in these patients, we have argued that "minimally conscious" does not adequately describe their residual cognitive abilities (Naci and Owen, 2013; Owen, 2013, Fernández-Espejo and Owen, 2013). Indeed, the patient described above was consistently and reliably able to communicate (using fMRI), which places him well beyond the diagnostic criteria describing the minimally conscious state. The term "functional locked-in syndrome" has also been proposed for patients who demonstrate consistent and reliable communication using solely adjunctive technologies (Giacino et al., 2009; Laureys and Schiff, 2012). In its classical clinical presentation, "locked-in syndrome" refers to patients who are left with only vertical eye movements and/or blinking, which often permits rudimentary communication.

Cognitive function, however, is generally fully preserved, at least in those cases where the lesion is limited to the ventral pons (Schnakers et al., 2008). Patients like the one described here are clearly "locked in" in the general sense of the term, but do not have many of the same neuropathological and clinical features of the classic locked in syndrome. Moreover, at present there is still considerable uncertainty about the full extent of residual cognitive function in such patients, and thus, about the suitability of the term "functional locked-in syndrome." That said, this is precisely the sort of question that can be explored with fMRI. Indeed, the patient has already been able to report that he remembers his own name and that he knows the current date and where he is (Naci and Owen, 2013), confirming that he is well oriented in time and space. He has also provided information about events that have occurred in the years since his accident, confirming that he is still able to encode new memories. Schnakers et al., (2008) have recently developed a standardized neuropsychological assessment for locked-in syndrome that uses simple eye movements as responses (in most cases to provide "yes"/"no" answers to questions). There is no technical or theoretical reason why a similar approach could not be used with fMRI data in entirely non-responsive patients, although the data would take considerably longer to acquire. To this end, Hampshire et al. (2013) have recently used fMRI to assess complex logical reasoning ability in a patient who was assumed to be in a vegetative state. Adapting a verbal reasoning paradigm from Baddeley (1968), Hampshire et al. (2013) presented participants with statements describing the ordering of two objects, a face and a house. Participants were instructed to deduce which of the objects was in front, and to visualize the object in their mind. For example, if they heard the statement "the face is not followed by a house," the correct answer would be "house." Conversely, if they heard "the face precedes the house," the correct answer would be "face." One patient, who based on the behavioral diagnosis was assumed to be in the vegetative state, engaged the same brain regions as healthy individuals in response to the reasoning task demands. This result was consistent with the patient's positive outcome in the fMRI command following task (Owen et al., 2006; Boly et al., 2007) and suggested that, despite the long standing clinical diagnosis of vegetative state, the patient was not only consciously aware, but, critically, retained capacity for higher-order cognition, in particular, for solving logically complex verbal problems.

Judicial Implications

The possibility of using fMRI or EEG for the detection of awareness in behaviorally non-responsive patients (Owen et al., 2006; Cruse et al., 2011; Naci and Owen, 2013; Fernández-Espejo and Owen, 2013) raises a number of issues for legal decision-making relating to the prolongation, or otherwise, of life after severe brain injury. At present, in most jurisdictions, decisions concerning life support (nutrition and hydration) are generally only taken once a diagnosis of permanent vegetative state has been made; that is, once one of the critical time thresholds described above has been reached. Until recently, fMRI or EEG had not demonstrated unequivocal signs of awareness in any patient that had survived beyond the time point required for such a diagnosis, but the case described in detail above is a vivid exception to that rule (Cruse et al., 2012b; Naci and Owen, 2013; Fernández-Espejo and Owen, 2013); that is to say, the patient has persisted in a condition of physical non-responsiveness for more than 12 years and has, therefore, long since met all of the internationally agreed criteria for a diagnosis of *permanent* vegetative state and could, therefore, be the subject of a legal petition to withdraw nutrition and hydration. Although this has not occurred in this particular case, we are aware of a number of cases that are currently being considered in various different legal jurisdictions, involving patients with similar clinical profiles. Typically, these cases involve one of two scenarios i) a dispute among family members; for example, the next of kin wishes to proceed with withdrawing nutrition and hydration, but other family members contest this decision on the grounds that it is not what the patient would have wanted (or what they want). ii) A dispute between medical staff and family members; for example, medical staff recommend withdrawing nutrition and hydration on the grounds of futility (the patient will never recover), but family members contest this opinion. In most of these cases, the key medical and legal decisions revolve around several inter-related factors: i) whether there is any chance of significant recovery; ii) whether the patient is conscious or "aware" of her/his condition; and iii) what the patient would have wanted if they could have been consulted about her/his current condition—in the latter case, an advanced directive or a "living will" is often used to guide the court's decision or, in the absence of such a document, the closest relatives are consulted and asked to evaluate what they think the patient would have wanted. Regarding the first of these factors, at present, there is no unequivocal evidence that the discovery of positive fMRI or EEG responses is predictive of recovery, although there are certainly some suggestions that this might be the case (Di et al., 2008; Coleman et al., 2009). Regarding the remaining two points of legal discussion, the case for the use of fMRI is becoming increasingly compelling. It is now clear that fMRI can be used to detect covert awareness in some cases where no clinical evidence exists to confirm that is the case (Owen et al., 2006; 2007; Monti/Vanhaudenhuyse et al., 2010; Cruse et al., 2011; 2012b; Naci and Owen, 2013; Fernández-Espejo and Owen, 2013), and, subject to the appropriate quality controls and scientific guidance, there is no a priori reason why such data could not be used to guide a court's opinion about "whether the patient is conscious or 'aware' of her/his condition." Again, the patient described in detail by Fernández-Espejo and Owen (2013) is a case in point; although multiple clinical assessments across twelve years suggested that he was "awake but unaware," the fact that he was able to report his own name, where he was, what year it was and whether or not he was in any pain, demonstrates beyond any doubt that he was "conscious" and "aware of his condition." More compellingly still, the fact that he could communicate, albeit in a rather rudimentary (fMRI) way, obviates any need for the court to consult the relatives about what the patient would have wanted and the need to locate, or rely upon, an advanced directive in reaching a decision. Ultimately, the morally challenging question of whether this is a life that is "*worth living*" (Kahane and Savulescu, 2009) is one that could be answered directly by the patient himself.

On the other hand, it is important to point out that neuroimaging of covert awareness is unlikely to influence legal proceedings where negative findings have been acquired. False-negative findings in functional neuroimaging studies are common, even in healthy volunteers, and they present particular difficulties in this patient population. For example, a patient may fall asleep during the scan or may not have properly heard or understood the task instructions, leading to an erroneous negative result. Indeed, in the recent study by Monti/Vanhaudenhuyse et al. (2010) no willful fMRI responses were observed in 19 of 23 patients—whether these are *true* negative findings (i.e. those 19 patients were indeed vegetative) or *false negative* findings (i.e. some of those patients were conscious, but this was not detected on the day of the scan) can not be determined. Accordingly, negative fMRI and EEG findings in patients should never be used as evidence for impaired cognitive function or lack of awareness.

CONCLUSIONS

In the last few years, neuroimaging methods—most notably fMRI and EEG—have been brought to bear on one of the most complex and challenging questions in clinical medicine, that of detecting residual cognitive function, and even covert awareness, in patients who have sustained severe brain injuries. The results demonstrate that responses need no longer be *physical* responses in the traditional sense (e.g. the blink of an eye or the squeezing of a hand), but can now include responses that occur entirely within the brain itself. The recent use of reproducible and robust task-dependent fMRI responses as a form of "communication" in patients who are assumed to be vegetative (e.g. Monti/Vanhaudenhuyse et al., 2010; Naci and Owen, 2013; Fernández-Espejo and Owen, 2013), represents an important milestone in this process. In some cases, these patients have been able to communicate information that was not known by the experimenters at the time, yet could be independently verified later (using more traditional methods of communication with the family), as being factually correct and true (Monti/Vanhaudenhuyse et al., 2010; Naci and Owen, 2013). More importantly perhaps, in one case, a patient has used these methods to answer clinically and therapeutically relevant questions (including "Are you in any pain?") that could not be answered in any other way, including via third party. The use of EEG—a more portable and cost effective method that can be used at the bedside—to detect of consciousness in patients who appeared to be entirely vegetative (Cruse et al., 2011; 2012a; 2012b), opens the door to the development and routine use of true "brain computer interfaces" in some of these patients (Owen, 2013).

One question that remains, both for neuroscience and for clinical practice, is where will this research lead us? There is no doubt that there currently exists a broad fascination, both among the general public and the media, about whether the methods described in this review could, and should, be used to ask patients whether or not they want to go on living. Although this is already a practical possibility, it is important to consider whether a simple "yes" or "no" response to such a question would be sufficient to ensure that a patient retained the necessary cognitive and emotional capacity to make such a complex decision. Clearly, it would not. Indeed, given the potential implications, if a robust and reliable response was obtained to such a question, one would want to be absolutely sure that the patient retained a level of decision-making capacity commensurate with the importance of any decision that might be made based on that response. In this context, decision-making capacity may be better considered as a continuum, rather than an "all or nothing" problem (Buchanan and Brock, 1989), with different thresholds required depending on the importance of the potential consequences of the decision. Clearly, decisions about the withdrawal of life support are of utmost importance, as they are radical and irreversible, and therefore, an appropriate level of decision-making capacity should be demonstrated before such a question could be even considered. Peterson and colleagues (2013) have recently laid out the conceptual foundations for a mechanistic explanation of capacity that would allow the necessary steps for incorporating neuroimaging data into the standard capacity assessment used in clinical practice. Clearly, we are entering an era where high-level assessments of residual cognitive function may soon be made based solely on fMRI (Hampshire et al., 2013) or EEG data, although a full assessment of the capacity for complex decision-making using any of the tools described in this review would still be extremely lengthy, logistically complex and, practically unfeasible in most contexts. Nevertheless, with the rapid emergence and deployment of so-called "brain-computer interfaces" for applications as diverse as gaming, the military, and coma, we would venture that within the next decade all of these obstacles will be overcome.

ACKNOWLEDGMENTS

This work was funded by generous awards to Adrian M. Owen from the Canada Excellence Research Chair (CERC) Program, the Canadian Institutes of Health Research (CIHR) and the Natural Sciences and Engineering Research Council of Canada (NSERC).

REFERENCES

Aguirre, G. K., Detre, J. A., Alsop, D. C., et al. (1996). The parahippocampus subserves topographical learning in man. *Cerebral Cortex, 6*, 823–829.

Andrews, K., Murphy, L., Munday, R., et al. (1996). Misdiagnosis of the vegetative state: Retrospective study in a rehabilitation unit. *BMJ, 313*, 13–16.

Bidet-Caulet, A., Fischer, C., Besle, J., Aguera, P. E., Giard, M. H., & Bertrand, O. (2007). Effects of selective attention on the electrophysiological representation of concurrent sounds in the human auditory cortex. *Journal of Neuroscience, 27*, 9252–9261.

Baddeley, A. D. (1968). A three-minute reasoning test based on grammatical transformation. *Psychometric Science, 10*, 341–342.

Birbaumer, N. (2006). Breaking the silence: Brain-computer interfaces (BCI) for communication and motor control. *Psychophysiology, 43*, 517–532.

Boly, M., Coleman, M. R., Davis, M. H., et al. (2007). When thoughts become action: An fMRI paradigm to study volitional brain activity in non-communicative brain injured patients. *NeuroImage, 36*, 979–992.

Buchanan, A. E., & Brock, D. W. (1989). Deciding for others. Cambridge, MA: Cambridge University Press.

Cerf, M., Thiruvengadam, N., Mormann, F., et al. (2010). *Nature, 467*, 1104–1108.

Childs, N. L., Mercer, W. N., & Childs, H. W. (1993). Accuracy of diagnosis of persistent vegetative state. *Neurology, 43*, 1465–1467.

Cincotti, F., Mattia, D., Babiloni, C., et al. (2003). The use of EEG modifications due to motor imagery for brain-computer interfaces. *IEEE Transactions on Neural Systems and Rehabilitation Engineering, 11*(2), 131–133.

Coleman, M. R., Davis, M. H., Rodd, J. M., et al. (2009). Towards the routine use of brain imaging to aid the clinical diagnosis of disorders of consciousness. *Brain, 132*, 2541–2552.

Cruse, D., Chennu, S., Chatelle, C., Bekinschtein, T. A., Fernández-Espejo, D., Pickard, J. D., . . . & Owen, A. M. (2011). Bedside detection of awareness in the vegetative state. *The Lancet, 389*(9809), 2099-2094.

Cruse, D., Chennu, S., Chatelle, C., Fernández-Espejo, D., Bekinschtein, T. A., Pickard, J. D., . . . & Owen, A. M. (2012a). The relationship between etiology and covert cognition in the minimally conscious state. *Neurology, 78*, 816-822.

Cruse, D., Chennu, S., Fernández-Espejo, D., Payne, W. L., Young, G. B., & Owen, A. M. (2012b). Detecting awareness in the vegetative state: Electroencephalographic evidence for attempted movements to command. *PLoS One*, *7*(11), e49933, 2012.

Di, H., Boly, M., Weng, X., et al. (2008). Neuroimaging activation studies in the vegetative state: Predictors of recovery? *Clinical Medicine*, *8*, 502–507.

Farwell, L. A., & Donchin, E. (1988). Talking off the top of your head: Toward a mental prosthesis utilizing event-related brain potentials. *Electroencephalography and Clinical Neurophysiology*, *70*, 510–523.

Fernández-Espejo, D., & Owen, A. M. (2013). Detecting awareness after severe brain injury, *Nature Reviews Neuroscience*, *14*(11), 801–809.

Fins, J. J., & Schiff, N. D. (2006). Shades of gray: New insights into the vegetative state. *The Hastings Center Report*, *36*, 8.

Gallivan, J. P., McLean, D. A., Valyear, K. F., Pettypiece, C. E., & Culham, J. C. (2011). Decoding action intentions from preparatory brain activity in human parieto-frontal networks. *Journal of Neuroscience*, *31*(26), 9599–9610.

Giacino, J. T. et al. (2002). The minimally conscious state: Definition and diagnostic criteria. *Neurology*, *58*, 349–353.

Giacino, J. T., Kalmar, K., & Whyte, J. (2004). The JFK Coma Recovery Scale-Revised: Measurement characteristics and diagnostic utility. *Archives of Physical Medicine and Rehabilitation*, *85*, 2020–2029.

Giacino, J. T., Schnakers, C., Rodriguez-Moreno, D., Kalmar, K., Schiff, N., & Hirsch, J. (2009). Behavioral assessment in patients with disorders of consciousness: Gold standard or fool's gold? *Progress in Brain Research*, *177*, 33–48.

Guger, C., Edlinger, G., Harkam, W., et al. (2003). How many people are able to operate an EEG-based brain-computer interface (BCI)? *IEEE Transactions on Neural Systems and Rehabilitation Engineering*, *11*, 145–147.

Hampshire, A., Parkin, B., Cusack, R., Fernández-Espejo, D., Allanson, J., Kamau, E., . . . & Owen, A. M. (2013). Assessing residual reasoning ability in overtly non-communicative patients using fMRI. *NeuroImage: Clinical*, *2*, 174–183.

Haynes, J. D., Sakai, K., Rees, G., et al. (2007). Reading hidden intentions in the human brain. *Current Biology*, *17*(4), 323–328.

Kahane, G., & Savulescu, J. Brain damage and the moral significance of consciousness. *Journal of Medical Philosophy*, *34*(1), 6–26.

Kotchoubey, B, et al. (2003). Cognitive processing in completely paralyzed patients with amyotrophic lateral sclerosis. *European Journal of Neurology*, *10*, 551–558.

Kulkarni, V. P., Lin, K., & Bendabis, S. R. (2007). EEG findings in the persistent vegetative state. *Journal of Clinical Neurophysiology*, *24*(6), 433–437.

Laureys, S., & Schiff, N. D. (2012). Coma and consciousness: Paradigms (re)framed by neuroimaging. *NeuroImage*, *61*, 478–491.

Monti, M. M., Coleman, M. R., & Owen, A. M. (2009). Executive functions in the absence of behavior: Functional imaging of the minimally conscious state. In S. Laureys, N. D. Schiff, & A. M. Owen (Eds.). *Coma science: Clinical and ethical implications. Progress in Brain Research*, Vol. 177. Oxford: Elsevier, pp 249–260.

Monti, M. M., Vanhaudenhuyse, A., Coleman, M. R., et al. (2010). Willful modulation of brain activity in disorders of consciousness. *New England Journal of Medicine*, *362*(7), 579–589.

Naccache, L. (2006). Psychology: Is she conscious? *Science*, *313*, 1395–1396.

Naci, L., Cusack, R., Jia, Z., V., & Owen, M. A. The brain's silent messenger: Using selective attention to decode human thought for brain-based communication. *Journal of Neuroscience, 33*(22), 9385–9393.

Naci, L., & Owen, A. M. (2013). Making every word count for vegetative patients. *JAMA Neurology, 70*, 1235-2041.

Owen, A. M., Coleman, M. R., Boly, M., Davis, M. H., Laureys, S., & Pickard, J. D. (2006). Detecting awareness in the vegetative state. *Science, 313*, 1402.

Owen, A. M., Coleman, M. R., Boly, M., Davis, M. H., Laureys, S., Jolles, D., & Pickard, J. S. (2007). Response to Comments on "Detecting awareness in the vegetative state." *Science, 315*, 1221.

Owen, A. M., & Coleman, M. (2008a). Functional imaging in the vegetative state. *Nature Reviews Neuroscience, 9*, 235–243.

Owen, A. M., & Coleman, M. R. (2008b). Detecting awareness in the vegetative state. In D. Pfaff (Ed.). *Molecular and biophysical mechanisms of arousal, alertness and attention. Annals of the New York Academy of Sciences*, 130–140. New York: Wiley-Blackwell.

Owen, A. M. (2013). Detecting consciousness: A role for neuroimaging. *Annual Review of Psychology, 64*, 109-1033.

Peterson, A., Naci, L., Weijer, C., Cruse, D., Fernández-Espejo, D., Graham, M., & Owen, A.M. (2013). Assessing decision making capacity in the behaviorally non-responsive patient with residual covert awareness. *AJOB Neuroscience, 4*(4), 3–14.

Pfurtscheller, G., & Neuper, C. (1997). Motor imagery activates primary sensorimotor area in humans. *Neuroscience Letters, 239*, 65–68.

Pfurtscheller, G., Scherer, R., Müller-Putz, G. R., & Lopes da Silva, F. H. (2008). Short-lived brain state after cued motor imagery in naive subjects. *European Journal of Neuroscience, 28*, 1419–1426.

Pfurtscheller, G., Brunner, C., Schlögl, A., & Lopes da Silva, F. H. (2006). Mu rhythm (de)synchronization and EEG single-trial classification of different motor imagery tasks. *NeuroImage, 31*, 153–159.

Royal College of Physicians. (2003). The vegetative state: Guidance on diagnosis and management. A report of a working party of the Royal College of Physicians. *Clinical Medicine, 3*, 249–254.

Schnakers, C. Majerus, S., Goldman, S., Boly, M., Van Eeckhout, P., Gay, S., . . . & Laureys, S. (2008). Cognitive function in the locked-in syndrome. *Journal of Neurology, 255*, 323–330.

Schnakers, C., Vanhaudenhuyse, A., Giacino, J., Ventura, M., Boly, M., Majerus, S., . . . & Laureys, S. (2009). Diagnostic accuracy of the vegetative and minimally conscious state: Clinical consensus versus standardized neurobehavioral assessment. *BMC Neurology, 9*, 35.

Sellers, E. W., & Donchin, E. (2006). A P300-based brain-computer interface: Initial tests by ALS patients. *Clinical Neurophysiology, 117*, 538–548.

Sorger, B., Dahmen, B., Reithler, J., Gosseries, O., Maudoux, A., Laureys, S., & Goebel, R. (2009). Another kind of "BOLD Response": Answering multiple-choice questions via online decoded single-trial brain signals. *Progress in Brain Research, 177*, 275–292.

Teasdale, G., & Jennett, B. (1974). Assessment of coma and impaired consciousness: A practical scale. *Lancet, 2*, 81–84.

The Multi-Society Task Force on Persistent Vegetative State. (1994). Medical aspects of the persistent vegetative state (I). *New England Journal of Medicine, 330*, 1499–1508.

Vaughan, T. M., McFarland, D. J., Schalk, G., Sarnacki, W. A., Krusienski, D. J., Sellers, E. W., & Wolpaw, J. R. (2006). The Wadsworth BCI Research and Development Program: At home with BCI. *IEEE Transactions on Neural Systems and Rehabilitation, 14*(2), 229–233.

Warren, M. A. (1973). On the moral and legal status of abortion. *The Monist, 57*(4), 1-9.

Weiskopf, N., Mathiak, K., Bock, S. W., Scharnowski, F., Veit, R., Grodd, W., . . . & Birbaumer, N. (2004). Principles of a brain-computer interface (BCI) based on real-time functional magnetic resonance imaging (fMRI). *IEEE Transactions on Biomedical Engineering, 51*, 966–970.

Wolpaw, J. R., McFarland, D. J., Neat, G. W., & Forneris, C. A. (1991). An EEG-based brain-computer interface for cursor control. *Electroencephalography and Clinical Neurophysiology, 78*(3), 252–259.

Wolpaw, J. R., Birbaumer, N., McFarland, D. J., Pfurtscheller, G., & Vaughan, T. M. (2002). Brain-computer interfaces for communication and control. *Clin Neurophysiol, 113*, 767–791.

Zeman, A. (2009). The problem of unreportable awareness. In S. Laureys, N. D. Schiff, & A. M. Owen (Eds.). *Coma science: Clinical and ethical implications. Progress in Brain Research*, Vol. 177. Oxford: Elsevier, pp 1–10.

8

Persistent Vegetative State, Akinetic Mutism, and Consciousness

WILL DAVIES AND NEIL LEVY

There is a strong intuition that conscious beings have a special kind—and perhaps an especially high degree—of moral significance denied to beings that lack consciousness. Although the way in which this special significance is spelled out is often misleading, the intuition is well grounded: Some kind of consciousness *does* make a significant difference to a being's moral status.

It is this intuition that explains the excitement surrounding recent work by clinicians and neuroscientists that apparently demonstrates consciousness in patients previously diagnosed as being in a persistent vegetative state (PVS).[1] We think that the inference from the experimental data to the presence of consciousness is too hasty.[2] Building on recent work by Colin Klein, we suggest that a key assumption underlying this inference is far less solid than it has appeared. Our argument casts doubt on methodologies that take *command following* as a marker for endogenous intentional agency and, hence, for consciousness. Klein himself nonetheless accepts that these patients are conscious, although for reasons other than their apparent ability to follow basic commands. We are not convinced by Klein's reasons for attributing consciousness, and therefore we adopt a more pessimistic position regarding the interpretation of the data. In closing, we argue that even if Klein is right and these patients do enjoy *some* kind of consciousness, this is not a kind of consciousness that could ground the special kind of moral status typically assigned to conscious human beings.

METHODOLOGICAL ASSUMPTIONS IN THE SCIENCE OF CONSCIOUSNESS

Our aim in this section is to review briefly the key findings that have motivated much of the discussion about borderline states of consciousness and then to highlight some key assumptions at work in the relevant scientific studies. We shall not rehearse the empirical evidence in any detail, given that other chapters in this volume provide extensive discussion. Briefly, there are two key pieces of evidence that have impressed commentators. Owen et al. (2006) showed that the neural activation in a PVS patient who was asked to imagine either playing tennis or navigating around her own house was very similar to the neural activation exhibited by healthy controls asked to perform the same tasks. Building on these results, Monti et al. (2010) used the same paradigm to develop what was, in effect, a communication system based on functional magnetic resonance imaging (fMRI), in which an apparently vegetative patient was able to answer "yes" or "no" to questions by imagining playing tennis or imagining navigating a familiar environment.

The precise details of the neural activation exhibited by PVS patients do not much matter for our purposes. What most interests us are the grounds for the inference from these experimental data to the conclusion that the patients are conscious. The key assumption underlying this inference, with which the experimenters and subsequent commentators largely seem to agree, is that the data provide evidence for the presence of *agency*, which in turn indicates the presence of consciousness (Bayne, 2013). More precisely, the patient's capacity to follow task instructions (e.g., to imagine playing tennis) is assumed to signal an endogenously generated intention to comply with the instructions, which in turn signals the presence of intentional agency and, hence, consciousness. We can break down this inference into two components:

> **Command Following**: Command following is a marker of intentional agency; that is, evidence for command following provides strong evidence for the presence of intentional agency.
> **Agency**: Intentional agency is a marker of consciousness; that is, evidence for intentional agency provides strong evidence for the presence of consciousness.

In combination, Command Following and Agency seem to guide much of the research into borderline states of consciousness.[3]

It has been said that first-person subjective report is the gold standard for the ascription of consciousness. In the absence of such reports, Agency surely provides the *silver* standard for consciousness. Because command following is taken to imply intentional agency, it in turn provides what we might think of as the *bronze* standard for consciousness:

> **Bronze Standard**: Command following is a marker of consciousness; that is, evidence for command following provides strong evidence for the presence of consciousness.[4]

Given that the patients at issue in these studies are incapable of verbal report, they do not satisfy the gold standard. Given that they do not exhibit signs of endogenous agency (despite apparent preservation of the neural and motoric machinery required for endogenous agency), they do not satisfy the silver standard. That leaves the bronze standard as the only option for researchers. Reliance on that standard is clearly visible in the work of Owen et al. (2006) and Monti et al. (2010).

In what follows, our aim is to cast doubt on the use of the bronze standard in such studies. While we do not dispute that this standard provides a useful means of studying consciousness in normal subjects, we have serious concerns about its application in severely brain-injured patients. Why accept that command following, a capacity to follow task instructions, entails the presence of intentional agency and, thence, consciousness?

COMMAND FOLLOWING AND AKINETIC MUTISM

Command Following looks rather shaky in the light of the following considerations. As recently discussed by Klein, the responsive PVS patient group bears many similarities to patients suffering from akinetic mutism (AM). AM is a wakeful state in which patients exhibit a severe and persistent decrease in responsiveness; a lack of spontaneous motor or verbal activity; indifference to pain, thirst, or hunger; flattened affect; and apathy without depression. As Klein interprets the data from AM patients, they are not capable of endogenous intention formation, and hence they lack a capacity for endogenous agency.[5] Despite their lack of spontaneous activity, AM patients are capable of following instructions and answering questions. Given the right prompts, AM patients may even engage in complex activity, such as reading a test and answering questions about it. Following Klein, we classify such responses as *stimulus-evoked cognition*, as contrasted with endogenously initiated intention and action. AM patients are claimed to be acting on commands without the mediation of endogenous intention and, hence, without endogenous agency. As such, Command Following simply does not stand up in AM.

The PVS patients in the cited studies and AM patients have damage to much the same parts of the brain.[6] In particular, both patient groups exhibit damage to the presupplementary motor area (SMA), which is heavily implicated in the literature on AM, and which many studies suggest is associated with voluntary, endogenous actions. Both patient groups exhibit a complete absence of self-initiated action or response, which is not explained by paralysis. Moreover, the studies of Owen et al. (2006) and Monti et al. (2010) suggest that these PVS patients likewise retain a latent capacity to follow task instructions and even answer questions.

Given the structural and behavioral similarities, then, it is plausible to suggest that responsive PVS patients are complying with the experimenters' requests in a manner that is analogous (or even identical) to the way in which AM patients engage in this behavior.[7] On this model, responsive PVS patients are exhibiting stimulus-evoked cognition in response to task commands but not an endogenously generated intention to imagine playing tennis or navigating about one's house. And if this is right, Command Following does not hold up in the study of borderline states of consciousness either.

One immediate response would be that although the command following exhibited by these patients is stimulus evoked and hence not endogenously initiated, it still signals a residual, functioning capacity for agency in these subjects. Even though these events are exogenously stimulated, the thought goes, they nonetheless establish the presence of an intentional response, and that is all that is required to establish agency and, hence, consciousness. Although this line of thought is tempting, we think it ought to be resisted. Suppose for the sake of argument that the data do establish the presence of an intentional response, by which we mean an event that is initiated and guided by an intention, an internal motivational state that prompts action (including mental action). We think that intentional responses in this sense do not suffice to establish *agency*. Agency is best understood as requiring the ability, not simply to act or respond, but to act in a way that is relatively independent of stimuli and that requires considerable flexibility of response. What is required is the formation and execution of not simply intentions but *endogenously initiated* intentions. The patient groups under discussion do not exhibit agency in this sense. The responsive PVS patients do not (as far as we know) issue responses of this sort without external prompting, and their responses are, moreover, extremely inflexible and constrained. Even granting the presence of intentions to act or imagine, command following does not suffice in this case for the attribution of agency, and hence does not suffice to establish the presence of consciousness.

Others are free to define "agency" as they wish, and one can imagine many less demanding notions than that sketched here. In particular, some may wish to tie agency to mere intentional response, even when that response is stimulus bound and inflexible. However, we think that the inference from agency, in this sense, to consciousness is far from compelling. At very most, it seems to us to provide grounds for ascribing consciousness to patients *only when they are exhibiting a stimulus-evoked intentional response.* Consciousness here would be "fleeting and stimulus-bound," as Klein puts it.[8] If agency is the operative marker for consciousness, and if agency is exhibited *only* when exogenously evoked and not at all otherwise, then presumably the evidential situation warrants the attribution of consciousness only when the subject is following the experimenters' commands. In contrast, agency in our preferred sense, which requires an endogenously initiated and flexible response, would provide much stronger evidence for a standing or uninterrupted conscious state.

So far, we have argued that Command Following looks very shaky in the case of patients with AM. Given the neural and behavioral similarities between AM patients and responsive PVS patients, it is plausible to model responsive PVS patients as sharing a deficit in endogenous intention formation. If this model holds, Command Following also looks shaky in these studies at the borders of consciousness, and this in turn entails that the responses of PVS patients do not provide strong evidence for the presence of consciousness. The PVS data can be interpreted as involving nothing more than stimulus-evoked cognition, hence no endogenous agency, and hence no consciousness.

As described in the next section, Klein nonetheless thinks that extreme AM patients and responsive PVS patients are in fact conscious. Given the foregoing arguments, clearly this inference has to have another basis besides Command Following and Agency. Our next task is to critically evaluate Klein's reasons for attributing consciousness in such cases.

ANOTHER ROUTE TO ATTRIBUTING CONSCIOUSNESS?

Despite casting doubt on Command Following, Klein nevertheless suggests that AM patients are conscious in some sense. It follows that he also sees no reason to deny consciousness to responsive PVS patients. In what sense are these patients deemed to be conscious? And on what grounds, if not via the bronze standard, are we supposed to attribute consciousness in such cases?

Regarding the first question, as to what type of consciousness is at issue, Klein suggests that AM patients possess *peripheral* consciousness without *focal* consciousness, a distinction derived from Kriegel (2004). As we understand it from Kriegel, peripheral consciousness is the sort of awareness that one has, for example, of objects in peripheral vision, the sound of passing cars as one listens to a concerto, or a background sense of cheerfulness resulting from a good mood. Focal consciousness, in contrast, is the sort of awareness one has of a foveated section of text held in front of one's face, or the melody line played by the soloist of a concerto. Klein seems to use "peripheral consciousness" in a somewhat looser sense: He views focal consciousness as involving the allocation of *attention*, whereas peripheral consciousness involves some awareness of *unattended* stimuli.[9] We think that nothing crucial hangs on this issue in what follows.

As for the second question, regarding the alternative route to attributing consciousness, Klein draws on two sources of evidence for the claim that AM patients are conscious (in the peripheral sense). The first source is self-reports by symptomatic AM patients. Patients with less severe forms of AM report a curiously "empty" mental state; these introspective reports are taken by Klein to be extremely good evidence that they are conscious. The second source is retrospective reports by patients who have recovered from AM. In some (but, importantly, not all) cases, these patients report having had some consciousness of their surroundings and of events when they were symptomatic.

We are skeptical that either source of evidence provides strong grounds for attributing consciousness to AM patients. Let us first consider self-reports by symptomatic AM patients. Here we should distinguish between two possible cases. The first is self-reports in patients with mild to moderate AM, who we assume are capable of *some* degree of spontaneous, self-initiated activity. The second is self-reports in patients with extreme cases of AM, who, let us assume, are completely incapable of any degree of spontaneous, self-initiated activity.

There are two worries concerning the first type of case. First, it is unclear why self-reports in such moderate cases of AM should be considered relevant to the argument concerning consciousness in PVS patients. PVS patients, we assume, are most similar to extreme AM patients, if indeed they are similar to AM patients at all. They are most akin to extreme AM patients because, like them, PVS patients exhibit no spontaneous, self-initiated activity whatsoever. As such, it is of questionable relevance that some moderate AM patients, who are capable of some self-initiated activity, self-report states of consciousness. A second and related worry is that if moderate AM patients are capable of some spontaneous activity, it is reasonable to infer that they retain some capacity for endogenous intention and agency. But if it is antecedently known that moderate AM patients have some capacity for endogenous agency, then self-reports of consciousness are not even essential: We can simply appeal to Agency, the claim that intentional agency is a strong marker for the presence of consciousness. (Recall that our target here has been Command Following, not Agency: For all we've said, Agency is just fine.) For moderate AM patients, then, we have an alternative, independent, basis for attributing consciousness—a basis that is unavailable in PVS patients, for whom spontaneous activity, and hence clear signs of endogenous agency, are crucially lacking. This disparity makes it wholly illegitimate to appeal to putative examples of consciousness in moderate AM patients to support claims of consciousness in responsive PVS patients.

We can also raise concerns regarding the second type of case, self-reports in patients with extreme AM.[10] As we have characterized them, extreme AM patients are completely incapable of any degree of spontaneous, self-initiated activity. As such, we can assume that any self-report issued by an extreme AM would have to be prompted or otherwise elicited. This creates a highly unusual situation regarding the evidential status of such reports. In order to understand this, let us briefly rehearse the dialectic to this point. Our overarching question is whether extreme AM patients, and likewise responsive PVS patients, are conscious. Both we and Klein agree that mere responsiveness to commands or stimulus-evoked cognition is insufficient to establish the presence of consciousness. Klein nonetheless believes that AM patients *are* conscious and is therefore required to provide some independent grounds for attributing consciousness in such cases. And this is the puzzle: What are we to say when the tendered grounds for attributing consciousness are stimulus-evoked self-reports regarding the patient's putative state of consciousness? Our inclination is to say that if stimulus-evoked cognition *in general* provides insufficient grounds to attribute consciousness, then we should also conclude that stimulus-evoked self-reports of consciousness provide insufficient grounds to attribute consciousness. We admit that there might be some superficial weirdness in this response, but it strikes us as the logical conclusion given the foregoing discussion.

We have been discussing Klein's first source of evidence for consciousness in AM: self-reports from symptomatic patients. Klein's second source of evidence is retrospective reports by patients who have recovered from AM. Some of the problems raised for self-report also apply to retrospective reports. In particular, if the retrospective reports are from someone who has recovered from mild or moderate AM, then they are of dubious relevance to the issue at hand. In order to focus on the most challenging case, however, let us consider a retrospective report from someone seemingly recovered from extreme AM.[11] Klein discusses a patient reported by Laplane, Baulac, Widlöcher, & Dubois (1984) who described his past state as like having "a blank in my mind." When quizzed on their private thoughts, a patient reported by Bogousslavsky et al. (1991) said, "I think of nothing. . . . I don't want anything." Another said he "did not have any projects for the future and did not have any personal thoughts."

Although statements of this sort are extremely interesting, it is unclear whether they provide any real evidence regarding the putatively conscious state of the individual while symptomatic. First, there are general worries about the quality and accuracy of testimony in such cases. Given the strikingly bizarre, uncommon, and elusive nature of extreme AM, perhaps it is simply not reasonable to expect recovered patients to have well-formed, reliable thoughts about their symptomatic state. Second, all of the quoted reports concern absences or failures, such as the absence of thought or desire. It is extremely unclear how to interpret such reported absences. One option would be to infer that these patient were, while symptomatic, in some sense conscious of having a "blank in my mind," and that this memory is reported in their retrospective appraisals. But another option, which seems at least equally plausible, is that the patient is just trying to find a way to report an absence of consciousness. These retrospective attempts might be aided or enhanced by the fact that the subject, while symptomatic, *ex hypothesi* would have been capable of certain types of stimulus-evoked cognition and hence would have retained some degree of informational sensitivity to sensory input. Recovered patients might even have access to some of the information that was taken up by the cognitive system during this period. However, recalling informational content that might have been conscious in a normal subject is not necessarily recalling consciousness. Recovered patients might also acquire some sort of awareness that their explicit beliefs, desires, and intentions did not change during the symptomatic period and might report this by saying that they "wanted nothing" or did not "have any projects" during this period. Such retrospective appraisals, however, clearly do not compel the attribution of conscious awareness of such absences while these patients were symptomatic.

Further analysis and careful reflection on these puzzling cases is certainly required. As things stand, however, we find no compelling reason to think that extreme AM patients—the most relevant group, as far as the comparison with PVS patients is concerned—are conscious. At the very least, if there *is* any consciousness in such patients, these states most likely would be fleeting and stimulus-bound.

THE MORAL STATUS QUESTION

Finally, we consider the moral significance of purported attributions of consciousness in PVS patients. Notwithstanding the skeptical conclusions of the previous two sections, let us suppose for the sake of argument that Klein is correct, that extreme AM patients are conscious, and that responsive PVS patients are similar enough in terms of their deficits to warrant the attribution of consciousness to them too. Let us even suppose that both sets of patients are conscious all (or much) of the time and not just when they are prompted. Would they then enjoy the moral status that is rightly attributed to normal subjects by virtue of the fact that they are conscious? We suggest that the answer is no.

One of us has previously argued that the bulk of the work in underwriting moral status is done not by the capacity for experience per se but by the capacity to have series of appropriately linked mental states (Levy, 2009; Levy & Savulescu, 2009). A being with an interest in having a life must be able to care how its life goes, and this requires very sophisticated cognitive abilities, such as an ability to conceive of oneself as a being persisting through time, to recall one's past, to plan, and to have preferences for how one's life goes (McMahan 2002; Singer 1993).

It seems clear that phenomenal consciousness is neither (metaphysically) necessary nor sufficient to sustain this kind of interest in having a life. Rather, it seems that what is required is for the being to self-attribute informational states of the right kind. This, in turn, requires the simultaneous availability of these informational states to (some of) the consuming systems constitutive of its mind. It follows that a necessary condition of possessing a serious interest in life is something akin to Block's (1995) notion of access consciousness. However, because we wish to avoid building in some of the commitments of Block's notion, such as the stipulation that information of which a being is access consciousness is poised for the rational guidance of its behavior, we shall call this kind of consciousness *informational consciousness*.

To have a serious interest in life, a being must self-attribute an appropriate set of its mental representations, where the contents of these representations concern its own plans and projects and the ground of these plans and projects: its own existence. A serious interest in life therefore requires self-consciousness, of this self-attributing kind, and this, in turn, depends on informational consciousness in the sense outlined earlier. Only a being that is self-conscious of an appropriate set of representations can hold itself as having plans and projects and therefore possess future-oriented desires with regard to these plans and projects. Only a being like this can suffer the distinctive harm associated with death: the permanent thwarting of these plans and projects. Only a being who can suffer such a harm has a reason to fear death (rather than dying).

Interpreting the moral significance of the foregoing experimental results therefore requires answering several questions, empirical and conceptual, in order to establish whether patients possess the capacity for informational consciousness (with the right contents) and for self-consciousness (again, with the right contents). As we have seen, Klein attributes *peripheral consciousness* to the PVS patient group. We have expressed doubts about this attribution, but for the purpose of this discussion we accept it. In the previous section, peripheral consciousness was characterized as the sort of awareness one has of items in the periphery of one's visual field, for example, or of a background mood of happiness. Perhaps this type of consciousness accompanies objects or states that are *unattended* by the subject. Given this understanding, one way to interpret states of peripheral consciousness is as a subset of states of phenomenal consciousness. The thought here is that if one's entire visual field constitutes one's total state of visual phenomenal consciousness, then presumably points in the periphery of this field would constitute a subset of states of visual phenomenal consciousness. As pointed out earlier, however, an exclusively phenomenal notion of consciousness such as this cannot do the work of underwriting a serious interest in life. On this reading, then, the attribution of peripheral consciousness gets us no closer to moral significance than the attribution of purely phenomenal consciousness.

In another interpretation of peripheral consciousness, this notion carries with it implications not (or not just) of phenomenal consciousness but of informational consciousness in the sense outlined previously. On this understanding, one's peripheral consciousness of a background mood of happiness, for example, would have informational content that is available to cognitive consuming systems. Even in this case, however, the prospects of underwriting a serious interest in life are not much improved. We assume that states of peripheral consciousness would have relatively thin, impoverished, indeterminate, and perhaps highly determinable informational contents. Peripheral consciousness therefore would offer very little to one's consuming systems in the way of specific, categorized, or recognitional information about one's environment, one's moods and desires, and so on. And this seems to fall well short of the sorts of content required for the formation, monitoring, and self-attribution of plans and projects by the subject. How could I so much as form, let alone self-attribute, a plan to reach for the glass by my bedside if my awareness of the glass is limited to the sort of vague, indeterminate, uncategorized awareness that one has of objects in the periphery of vision?

There are many complications here that warrant more extended discussion. One such complication is that insofar as responsive PVS patients could be said to have *some* kind of consciousness, they would admittedly have a degree of moral status that completely nonconscious beings lack. More specifically, if these patients possess peripheral consciousness, we assume that they could have some peripheral awareness of pain and pleasure. We are somewhat unclear on what the nature of such peripheral pains and pleasures would be: For example, would the attention-grabbing nature of pain put pressure on the idea that one could ever be *merely* peripherally conscious of pain? Whatever the truth of the matter, the mere peripheral awareness of pain and pleasure might create obligations to take these states into account, such as issuing analgesics or creating environments in which sensory pleasures can be increased.

A second complication concerns the suggestion (made by Klein) that subjects who possess a *standing* state of peripheral consciousness might come to have states of focal consciousness when prompted. That is to say, perhaps issuing a command to an unresponsive patient could shift them from an indeterminate and inattentive state of awareness to a state with richer content through greater allocation of attention. If this were the case, it would be reasonable to think that they could come to have focal consciousness given other kinds of stimuli. In particular, they might have full-blown focal awareness of pain and pleasure, states that are plausibly of direct moral significance. Here the moral obligations to take these states into account in making treatment decisions, for example, would be much clearer-cut than in the case of merely peripheral pain and pleasure. Even granting these points, however, one of us thinks that these experiences in normal subjects get a great deal of their moral significance from their wider role in agents' mental lives, not from their raw phenomenal feel (Levy, 2009). Unless it could be established that these patients are also deploying the resources of informational consciousness and self-consciousness in registering and self-ascribing these states, then it follows that pain is not as bad and pleasure not as good for these patients.

CONCLUSIONS

The mental life of patients in extreme and prolonged states of unresponsiveness remains one of the most puzzling and ethically significant issues in contemporary brain science and medicine. We believe that the waves of excitement surrounding putative findings of consciousness in PVS patients have been premature. The methodology encapsulated in the bronze standard, which takes the ability to follow commands as a proxy for consciousness, is of dubious standing. Following a plausible model of the impairments characterizing AM, command following cannot be adopted as a straightforward marker of endogenous intentional agency in such patients. The mere capacity for stimulus-evoked cognition and response, in our view, does not suffice for agency and hence cannot provide the desired link to consciousness. Much further work is clearly needed to understand the range and limits of exogenously initiated intention and action. Such future work will shed further light on the mental states of patients suffering these profound impairments in the capacity for endogenous action. Perhaps at some later stage, another route to the attribution of consciousness will be found in such patients. We would welcome such developments. As we have argued, however, even granting such claims of consciousness, their ethical significance should not be immediately inflated. One plausible candidate for the putative consciousness exhibited in such cases, peripheral consciousness, lacks the properties required to ground serious self-interest and, hence, morally significant agency.

NOTES

1. It is unclear to us how the putative cases of consciousness should alter our classification of these patients. It might seem that such consciousness would shift these patients from a classification of PVS to the minimally conscious state (MCS). Another possibility, however, is that the degree of consciousness would be sufficient to place these patients in the category of locked-in syndrome. We remain neutral on these classificatory issues.

2. One of us has previously questioned whether the data show that the patients have the kind of consciousness routinely assumed to be at issue, namely phenomenal consciousness. Because phenomenal consciousness is not the kind of consciousness that actually does the bulk of the work in underwriting moral status of the kind and degree at issue, however, that quibble now seems irrelevant.
3. Indeed, one of the core diagnostic criteria for transitioning into MCS from PVS is "following simple commands" on a "reproducible or sustained basis" (Giacino et al. [2002], p. 351).
4. Because command following is normally taken to *imply* intentional agency, whereas agency *signifies* consciousness, the link between command following and agency is strictly stronger than the link between agency and the attribution of consciousness.
5. Klein adopts a scientific notion of "intention," which "simply stands for whatever internal motivational state gives rise to a particular action, subject only to the restriction that it is a sufficiently complex state that it cannot occur completely automatically."
6. For more details see Klein §3.3.1.
7. Note that the claim is not that responsive PVS patients *are* AM, just that there is a shared deficit of endogenous intention and, hence, of endogenous agency. The explanatory power of such a model should be judged on its merits, then, not on the plausibility of a diagnosis of AM in the PVS patient group.
8. Klein considers and rejects this interpretation of "fleeting" consciousness, which has been proposed elsewhere in the literature by Damasio (2000).
9. Klein has endorsed our interpretation in correspondence.
10. We are not in fact sure whether any such reports exist, but this does not matter. The argument goes through all the same.
11. Again, we are not sure whether such retrospective reports from genuine *extreme* AM patients exist. This does not matter for our argument.

REFERENCES

Bayne, T. (2013). Agency as a marker of consciousness. In T. Vierkant, J. Kiverstein, and A. Clark (Eds.) *Decomposing the will* (pp.160–180). Oxford: Oxford University Press.

Block, N. (1995). On a confusion about a function of consciousness. *Behavioral and Brain Sciences, 18*, 227–287.

Bogousslavsky, J., Regli, F., Delaloye, B., Delaloye-Bischof, A., Assal, G., & Uske, A. (1991). Loss of psychic self-activation with bithalamic infarction. *Acta Neurologica Scandinavica, 83*, 309–316.

Damasio, A. (2000). *The feeling of what happens: Body and emotion in the making of consciousness.* New York: Harcourt, Brace, and World.

Giacino, J. T., Ashwal, S., Childs, N., Cranford, R., Jennett, B., Katz, D. I., . . . & Zasler, N. D. (2002). The minimally conscious state: Definition and diagnostic criteria. *Neurology, 58*, 349–353.

Klein, C. (2015). Consciousness, intention, and command-following in the vegetative state. *British Journal for the Philosophy of Science*. doi: 10.1093/bjps/axv012

Kriegel, U. (2004). The functional role of consciousness: A phenomenological approach. *Phenomenology and the Cognitive Sciences*, *3*, 171–193.

Laplane, D., Baulac, M., Widlöcher, D., & Dubois, B. (1984). Pure psychic akinesia with bilateral lesions of basal ganglia. *Journal of Neurology, Neurosurgery, and Psychiatry*, *47*, 377–385.

Levy, N. (2009). What difference does consciousness make? *Monash Bioethics Review*, *28*, 1–13.

Levy, N., & Savulescu, J. (2009). The moral significance of phenomenal consciousness. *Progress in Brain Research*, *117*, 361–370.

McMahan, J. (2002). *The ethics of killing: Problems at the margins of life*. Oxford: Oxford University Press.

Monti, M. M., Vanhaudenhuyse, A., Coleman, M. R., Boly, M., Pickard, J. D., Tshibanda, L., . . . & Laureys, S. (2010). Willful modulation of brain activity in disorders of consciousness. *New England Journal of Medicine*, *362*, 579–589.

Owen, A. M., Coleman, M. R., Boly, M., Davis, M. H., Laureys, S., &. Pickard, J. D. (2006). Detecting awareness in the vegetative state. *Science*, *313*(5792), 1402.

Singer, P. (1993). *Practical ethics*. Cambridge: Cambridge University Press.

PART III

Ethics

9

Lay Attitudes to Withdrawal of Treatment in Disorders of Consciousness and Their Normative Significance

JACOB GIPSON, GUY KAHANE, AND JULIAN SAVULESCU

With the growing sophistication of modern medical interventions, patients are increasingly being kept alive in states of severely diminished consciousness after injuries that in the past would have been fatal (Bernat, 2006; Jennett & Plum, 1972). This has led to a number of patient outcomes not previously seen, necessitating the development of a more nuanced approach to classification and diagnosis of patients with consciousness below typical levels. Collectively, these are known as the disorders of consciousness. They include coma, vegetative state (VS), and minimally conscious state (MCS). A relevant state that is clinically related, although consciousness is preserved, is locked-in syndrome (LIS). Patients in these conditions may be kept alive almost indefinitely by artificial nutrition and hydration, necessitating decisions on whether sustaining life is morally appropriate and judgments on which lives are worth living. Discussion of the morality of withdrawal of treatment has been rife in medical, legal, and ethical contexts, and the issue remains highly controversial.

One issue that has been largely ignored in this debate has been the actual views and opinions of laypeople who may one day find themselves in a situation such as this. Although lay attitudes cannot decide moral issues, they have significant implications for policy development and may even have some normative significance in and of themselves.

Inquiries into lay attitudes have become increasingly prominent in recent ethical discourse and are often termed "empirical ethics." Broadly speaking, this approach employs qualitative and quantitative social science methods not traditionally used in ethical research, in conjunction with normative discussion, to add to moral understanding. This chapter applies such an approach to the question of the morality of withdrawal of treatment from people with disorders of consciousness.

We begin with a brief review of these disorders and of the ethical considerations that have been regarded as relevant to the question of whether life-sustaining treatment should be continued or withdrawn. We then briefly summarize the findings of prior surveys that have largely focused on the views of medical practitioners before reporting in some detail the results of a survey (Gipson, Kahane, & Savulescu, 2014) we conducted into lay attitudes regarding these questions. We conclude by discussing the potential normative significance of our findings.

MEDICAL BACKGROUND

The term *vegetative state* (VS) was first coined in 1972 to describe patients who demonstrated normal sleep-wake cycles but no evidence of awareness of either themselves or their environment. In this condition, no motor, sensory or visual function is observed beyond reflexes (Jennett & Plum, 1972). Studies performed thus far strongly suggest these patients lack conscious experience and therefore cannot feel pain (Jennett, 2002; Royal College of Physicians, 1996; Zeman, 2001). Despite some controversy in the nomenclature, VS is commonly termed "permanent" if it persists for longer than 1 year after traumatic injury or 3 to 6 months after anoxic events. Because of reports of extremely rare late-term recovery to higher states of consciousness, however, this qualifier is often dropped.

MCS is similar to VS but with partial preservation of awareness at drastically reduced levels. Patients may inconsistently be capable of simple command following, "yes/no" responses, verbalization, or purposeful behavior. Brain activation similar to that of healthy controls in response to painful (Boly et al., 2005, 2008), and emotional (Bekinschtein et al., 2004) stimuli has been demonstrated, although exact levels of consciousness are unknown.

LIS was first described in 1966 (Plum & Posner, 1972). It is a disorder of movement and responsiveness rather than of consciousness. Patients retain normal cognition but are completely paralyzed and unable to speak or move. Most are able to communicate using eye movements, although some lack even this capacity; this is termed total LIS (American Congress of Rehabilitation Medicine, 1995). This condition is included in discussions because of its similar clinical presentation to disorders of consciousness. Although it is described mainly as a speculative risk, there is a theoretical possibility, given the reliance on response to stimuli when diagnosing disordered consciousness, that any patient diagnosed as being in VS or MCS may retain awareness and may in fact suffer from total LIS.

ETHICAL PRINCIPLES

The ethical basis for decisions surrounding withdrawal of treatment from these conditions relies on a number of well-described moral principles. The four most prominent of these in contemporary bioethics are the principles set out by Beauchamp and Childress (1994): *autonomy, beneficence, nonmaleficence,* and *justice.*

Autonomy—the principle that people have sovereignty over their own body—is often give preeminence in Western ethical and legal contexts. Appeals to autonomy are hampered in this context, however, given that patients in VS do not appear to have the cognitive capacities to form desires of any kind and it is therefore doubtful that they are capable of forming autonomous desires. If these desires exist, patients are in any case unable to communicate them. A common compromise is to uphold the desires of the previously competent patient. Advance directives and the use of surrogate decision makers are devices designed to allow this. One worry about this approach, however, is that individuals lack both the understanding and the imagination to comprehend conditions of such severe brain damage, raising doubts about the appropriateness of letting their present preferences influence their future medical treatment (Buchanan & Brock, 1989; Savulescu, 1994).

Moreover, the moral basis of relying on the preferences of a previously competent patient has also been challenged at a more fundamental level. Several authors advocate the view that there is a discontinuity between a patient's previous interests while competent and current interests while in a state of diminished consciousness (Parfit, 1987; Robertson, 1991). Relying on previously expressed wishes therefore binds an individual in VS or MCS to previously held values and desires that no longer have any relevance or authority. Because many people with LIS are able to communicate, ascertaining their desires is less problematic, and requests for withdrawal of treatment or euthanasia are uncommon (Doble, Haig, Anderson, & Katz, 2003).

Beneficence (do good) and nonmaleficence (do no harm) are often amalgamated in appeals to the best interests of these patients. However, it is controversial whether patients in VS or even MCS can even be said to have interests, given their limited cognitive capacities. If they do have interests, it is also highly controversial whether these interests are really served by continued treatment. For example, the continuation of an insensate, insentient, and, arguably, undignified life may be thought to be an objectively bad existence and thus against a person's interests. However, some argue, in a modern version of Pascal's wager, that given VS patients are unaware, being kept alive cannot hurt them. Therefore, given the possibility of recovery, as unlikely as it may be, it will always be in a patient's best interests to have treatment continued just in case recovery does occur (Stone, 2007). It has often been assumed that higher levels of cognitive capacity and consciousness in the MCS gives these patients a greater interest in continued treatment. It has been argued, however, that given the potential for MCS patients to feel pain and the possibility of having insight into their plight, such patients are in fact in a worse situation than those in VS, and for that reason, there is greater moral impetus to withdraw treatment from patients in MCS (Ashwal & Cranford, 2002; Johnson, 2011; Kahane & Savulescu, 2009; Nelson & Cranford, 1999). Notwithstanding their significant functional deficits, the majority of people with LIS report good quality of life (Lulé et al., 2009). A consideration of the best interests of these patients would therefore support continued treatment. Surveys of quality of life are, however, restricted to those in LIS who are able to communicate. It is likely that, if able, people in total LIS would report a worse quality of life.

Distributive justice is the principle of allocating the benefits of medical treatment in a fair way given the finite nature of the health budget. Despite its status as one of the four main principles in bioethics, considerations of distributive justice are rarely invoked in the literature surrounding this issue, possibly because of the perceived callousness of allowing finances to dictate life-and-death situations. The costs are, however, substantial. Treatment involves the indefinite provision of expensive care with uncertain gain. A recent research paper produced by the University of York estimated the total yearly cost of caring for a person in VS as approximately £91,500 (Formby et al., 2015). No equivalent estimate of the cost of caring for a person in MCS could be found, but costs are likely to be broadly similar, depending on the level of function in a particular patient.

The sanctity of life is another putative moral principle that is often invoked in connection with debates about withdrawal of life-sustaining treatment in such patients. In this view, which is often associated with certain religious tenets, particularly those of the Roman Catholic church (John Paul II, 2004), life is sacred and it is impermissible to terminate the life of a person, regardless of whether continued life is not in the person's interests or even contrary to them, and regardless of whether the person wishes to end their life—let alone because resources are scarce.

The nebulous concept of "dignity" is also often invoked by both supporters and opponents of withdrawal of treatment, despite being neither well defined nor understood (Schroeder, 2010). Supporters of withdrawal of treatment argue that being dependent on medical interventions such as feeding tubes reduces a person's dignity (Andrews, 1997), whereas opponents invoke a view of dignity as an intrinsic property of human beings that cannot be diminished (John Paul II, 2004).

PREVIOUS SURVEYS

A number of surveys have assessed the levels of support for allowing withdrawal of treatment. These have predominately focused on the views of healthcare professionals in the case of VS and have demonstrated levels of support ranging from 66% to 89% among physicians from the United States and Europe (Demertzi et al., 2011; Hodges, Tolle, Stocking, & Cassel, 1994; Payne, Taylor, Stocking, & Sachs, 1996). However, the proportion in favor varied with the population surveyed.

Religion and culture are likely to explain some of this variation, with physicians from Western countries such as Australia and Northern Europe generally more likely to be willing to withdraw artificial nutrition and hydration than other cultures. For example, a study of Japanese physicians found that only 17% would withdraw treatment from a VS patient even if the family and previously expressed wishes were in agreement that treatment should be withdrawn (Asai et al., 1999). A similar study in Mexico found only 48% support among physicians for this position (Lisker, Alvarez Del Rio, Villa, & Carevale, 2008).

Studies have also suggested that age and greater religious conviction correlate with a greater reluctance of physicians to withdraw artificial nutrition and hydration from VS patients (Vincent, 1999). This finding parallels the results of studies in intensive care units showing that older, more religious physicians are less likely than others to accept treatment limitations (Christakis & Asch, 1995; Sprung et al., 2003).

Interestingly, several studies have found that a much higher percentage of physicians would want treatment withdrawn from themselves than would agree to its appropriateness in others. In one study, only 66% of medical and paramedical professionals favored withdrawal of treatment from patients in chronic VS, but 82% of respondents admitted they would not want to live that way (Demertzi et al., 2011). One study found that 29% of neurologists would favor receiving less intensive life-sustaining treatment than they themselves would give a patient (Kuehlmeyer et al., 2012).

As an illustration of this phenomenon, consider the following excerpt from a qualitative study of how caregivers of patients in VS made decisions about withdrawal of treatment (Kuehlmeyer, Borasio, & Jox, 2012). A woman caring for her husband with VS states that, even though she believes it is "inhumane" to withdraw treatment from her husband, she would want treatment withdrawn if she were in that condition:

> Respondent: Unplug everything.
> Interviewer: Everything?
> Respondent: Everything.
> Interviewer: Including the artificial nutrition?
> Respondent: Yeah, everything. Well I wouldn't want to live like that. Mm Mm [negation]. Before, you haven't made up your mind. But if you experience something like that for yourself, then no.
> Interviewer: But you've just said that starvation is inhumane.
> Respondent: Yes, for others. For myself, phh [non-verbal expression, to say I don't mind].

The same study seemed to indicate that decisions about treatment withdrawal were rarely based on patient autonomy. Even when caregivers had information about patients' previously expressed wishes in the form of advance directives, these were frequently ignored, and the caregiver opted to continue treatment nonetheless.

This refusal to advocate irreversible decisions for others that caregivers and doctors would want done to themselves may reflect a desire to, above all, "do no harm," given the uncertainty about whether patients can feel pain, the ambiguity of prognosis, and so on. Commenting on a similar phenomenon in a European survey of withdrawal of treatment from MCS and VS patients, Demertzi et al. (2009) suggested that survey respondents may have "draw[n] a line between expressing preferences for self versus others, by implicitly recognising that the latter could be a step on the slippery slope to legalise euthanasia."

Compared with VS, treatment withdrawal from patients in MCS enjoys far less support, although much less research has been done in this area. One study of healthcare professionals found that 28% supported withdrawal of treatment in MCS compared with 66% in the case of VS (Demertzi et al., 2011).

Given the potential for beliefs about whether these patients can feel pain to influence beliefs about the appropriateness of withdrawal of treatment, it is interesting to assess the preponderance of these beliefs. Twenty years ago, a national survey of U.S. medical directors and neurologists (Payne et al., 1996) found that 30% believed VS patients could feel pain, despite this being incompatible with clinical guidelines. A more recent study by Demertzi et al. (2009) showed that 56% of European doctors believed patients in VS could feel pain. Paramedical professionals, people with religious convictions, and older caregivers were more likely than others to believe VS patients could feel pain. A much higher proportion of doctors (96%) believed patients in MCS could feel pain. Importantly, the belief that patients can feel pain has shown an inverse correlation with support for withdrawal of treatment (Demertzi et al., 2013). The authors hypothesized that the belief that patients can feel pain indicates that these doctors ascribed conscious experiences to the patient, along with its attendant moral significance.

Few data are available on attitudes toward treatment withdrawal in cases of LIS. As mentioned earlier, people with LIS generally report good quality of life and would reject having their treatment withdrawn. Despite this, there is concerning evidence to suggest that clinicians involved in their care may support less aggressive treatment in LIS patients due to false perceptions of reduced quality of life (Laureys et al., 2005).

A SURVEY OF LAY ATTITUDES

As previously stated, the views and attitudes of laypeople regarding these questions have been conspicuously absent from the debate. It is still unclear what proportion of laypersons favor treatment withdrawal in such conditions, nor is it clear what moral values and principles shape these views and whether these values are similar to those that have dominated the debate within ethics. Similarly, it is unclear whether the trends found among medical professionals—the propensity for older, more religious physicians to reject treatment withdrawal; a greater desire to have treatment withdrawn from themselves than to advocate for it for others; and a greater level of agreement with treatment withdrawal from VS compared with MCS—are also be present in laypeople.

In a survey designed to begin addressing these questions, we asked 199 U.S. residents their views on withdrawal of treatment from patients in VS, MCS, LIS, and total LIS using an anonymous online survey (for full details, see Gipson et al., 2014). In addition, we asked how important principles such as autonomy, best interests, distributive justice, sanctity of life, and respect for human dignity were in making these decisions. In order to investigate lay views about a concrete actual case, participants were also asked whether they agreed with a British court's decision to reject withdrawal of treatment from a woman in MCS (*W v. M*, 2011). We also asked participants to respond to a widely studied moral dilemma, known as the footbridge dilemma (Gold, Pulford, & Colman, 2014), in order to assess participants' general moral views to severely harming a single individual in order to benefit a greater number. Finally, participants were asked to complete three independently validated, commonly used measures of personality, cognitive tendencies, and empathic concern (the Big Five Inventory, the Need for Cognition Scale, and the Interpersonal Reactivity Index) in order to investigate possible relationships between a person's ethical beliefs and underlying psychological traits.

Lay Attitudes to Withdrawal of Treatment

Lay attitudes to withdrawal of treatment were measured first by asking participants to rate their agreement with the statement, "It is morally acceptable to end the patient's life by stopping treatment in [each of the conditions]." Level of agreement was assessed with a seven-point Likert scale, with 1 correlating to strongly agree and 7 to strongly disagree. This seven-point scale was simplified to a categorical answer by assigning a score of 1 to 2 as agreement, 6 to 7 as disagreement, and 3 to 5 as unsure or having no strong opinion in order to determine the prevalence of definite agreement with treatment withdrawal from each of the conditions. Responses are shown in Table 9.1.

We found that support for withdrawal of treatment from patients in VS or total LIS was higher than for those in MCS or LIS. Our findings may at first appear discordant with those of some previous surveys, which have shown a much higher level of support for withdrawal of treatment from VS than the 40.2% shown here. However, this difference is likely to be due largely to methodological differences, which make comparisons difficult. Previous surveys have relied on a forced yes/no dichotomy, whereas our study asked participants to indicate their view on a continuous scale, giving them space to indicate uncertainty. One striking finding of our survey is that when people are given this option, a large proportion select it in each case. Clearly, there is a lot of moral uncertainty that has been somewhat obscured by the methodology of prior studies. If we exclude the 42.2% of people who had no strong opinion either way, 69.6% of remaining participants agreed to some extent with treatment withdrawal in VS—a rate that is much closer to the levels found by previous surveys.

Another important difference is that earlier studies were focused on medical professionals, who presumably have much greater familiarity with these issues than the laypeople surveyed here. This difference likely translated into greater acceptance of treatment withdrawal. A previous study that included both physicians and paramedical professionals found a lower level of 66% in agreement

Table 9.1 Response to the Statement, "It is Morally Acceptable to End a Patient's Life by Stopping Treatment in [Each of the Conditions]."

	Agree	**Unsure**	**Disagree**
VS	40.2%	42.2%	17.6%
MCS	20.6%	38.2%	41.2%
LIS	25.3%	38.9%	35.8%
Total LIS	35.2%	38.2%	26.6%

(Demertzi et al., 2011). Greater unwillingness to withdraw treatment as we move from physicians to laypeople does therefore seems to be a consistent finding.

In line with previous studies, agreement with withdrawal of treatment from patients in MCS was much weaker than for VS. The higher level of functioning in MCS as opposed to VS appears to be a strong barrier to endorsing withdrawal of treatment. Whereas many people could be quite sure that removal of treatment from VS patients is ethically permissible, the presence of consciousness in MCS signifies the presence of interests and a life that might be deprived by withdrawing treatment. Given that these decisions must be made in conditions of uncertainty and the consequences involve loss of life, it may seem reasonable to many people, even those who are in favor of treatment withdrawal in the case of VS, to adopt a risk-averse position and disapprove of it in the case of MCS, at least until a stronger ethical consensus is reached regarding the moral status of such an intervention. By contrast, there is greater liberal consensus, highlighted in several high-profile cases, that withdrawal of treatment in VS is at least sometimes appropriate.

When asked to rank the four conditions in terms of which was worse for the patient, participants rated MCS as better than VS. Although this result is unsurprising, given that a lower percentage of respondents supported treatment withdrawal in MCS, it does contradict the views of some ethicists who have argued that because MCS patients may be able to feel pain and possibly have greater insight into their condition, they may be actually in a worse position than VS patients (Johnson, 2011; Kahane & Savulescu, 2009; Nelson & Cranford, 1999). One author wrote that, "just as VS is considered to be 'a fate worse than death,' being in a permanent MCS is a fate worse than VS" (Ashwal & Cranford, 2002).

In Table 9.2, we report participants' attitudes to withdrawal of treatment when they were asked to consider this question in the first person. Consistent with prior surveys of medical practitioners, we found greater willingness to condone withdrawal of treatment when this scenario was considered from the first-person, rather than the third-person, perspective. Past observers

Table 9.2 Response to the Statement, "I would want Treatment Withdrawn if I were in [Each of the Conditions]."

	Agree	Unsure	Disagree
VS	64.2%	21.7%	14.1%
MCS	41.4%	36.4%	22.2%
LIS	35.8%	38.9%	25.3%
Total LIS	55.8%	27.1%	17.1%

of this effect have hypothesized that this difference may be attributed to legal ambiguity surrounding end-of-life care for brain-damaged patients (Demertzi et al., 2011). It has been suggested that many medical practitioners, while largely accepting the permissibility of treatment removal in VS, may nevertheless worry that allowing it generally would put policy on a slippery slope toward legal euthanasia. Another possibility is that participants recognize that decisions that have permanent and drastic consequences for the individual should be autonomous, and it also may be an indication that the current safeguards in place to ensure this are seen to be inadequate.

It is also possible that this difference is in part due to framing effects. Participants were first asked whether they found treatment withdrawal to be ethically acceptable, and only then whether they would want this in their own case. The latter question is less abstract and more vivid, and participants may have reached different moral conclusions once they imagined the situation from the first-person perspective. It is also possible that when participants were asked whether withdrawal of treatment is ethically acceptable, some thought they were being asked whether treatment should be withdrawn in *all* cases as a general rule, not whether there are circumstances in which this might be acceptable (e.g., when withdrawal was in accordance with a patient's previously expressed wishes).

In any event, it is unlikely that participants both wanted treatment withdrawn in their own case *and* thought that withdrawal is generally morally wrong even when such wishes have been expressed. Therefore, we can conclude that laypeople accept the permissibility of withdrawal of treatment in disorders of consciousness to much higher levels than are revealed by direct questioning, although they may still have reservations about accepting this conclusion as a general policy.

W v. M

Over the past decades, the legality of withdrawal of treatment from patients in chronic VS has been confirmed in several jurisdictions in a number of high-profile cases, including in the United States (*Cruzan v. Director, Missouri Department of Health*, 1990), United Kingdom (*Airedale NHS Trust v. Bland*, 1993), and Australia (*Gardner: re BWV*, 2003), although the same has not been the case for MCS. One important exception is a 2011 case from the United Kingdom, *W v. M*, in which a family's application to have treatment withdrawn from a woman in MCS, known only as M, was denied. The family believed that M would not have wanted to live this way and relied on comments she made while still competent saying she did not want "a life dependent on others" and would prefer to "go quickly." The judge

Table 9.3 Response to Question, "Should Treatment be Stopped [in the Case of *W v. M*]?"

Agree	47.0%
Unsure	30.3%
Disagree	22.7%

in this case, however, gave significant weight to the presence of consciousness in this woman, even if at a greatly reduced level, and therefore ruled that it would not be in M's best interests to have treatment withdrawn.

One aim of our study was to investigate whether this ruling is in line with the moral views of laypeople (Table 9.3). Despite that fact that only 20.6% of people agreed it is morally acceptable to end a patient's life by stopping treatment in MCS when that condition was described to them in general terms, 47.0% reported that this should have happened in the case of *W v. M*. Indeed, it is striking that only 22.7% explicitly disagreed with treatment withdrawal. Thus, there was approximately double the support for withdrawal than there was opposition. People were more in favor than against withdrawal in this specific case, contrary to the judgment.

This result suggests that when laypeople considered concrete cases of MCS, they were far more willing to endorse treatment withdrawal than when they considered MCS in abstract, general terms. Another plausible explanation for this result is that there is good evidence that the patient, while in her competent state, expressed a desire to have treatment withdrawn if she ever lost her capacity for consciousness. We may therefore conclude that although people may oppose treatment withdrawal from MCS as a general rule, they nevertheless see withdrawal of treatment as justified when a person's wishes support it. Although laypeople may be reticent about removing treatment, the importance given to the patient's autonomy may override these considerations in cases in which a patient's previous wishes are known. One important implication of this result is that laypeople may accept the permissibility of treatment withdrawal in certain situations to higher levels than is revealed by direct questioning, something that may have led to an underestimation of support for treatment withdrawal from patients in MCS in previous studies.

Factors Underlying Decision Making

In order to gain an understanding, not just of the views of laypeople regarding withdrawal of treatment, but also of the moral considerations that underlie these views, we asked participants to also report how religious they were and the importance they ascribed to a number of morally relevant factors.

We found that 54% of participants identified with a denomination of Christianity, 40% with no particular religion, 4% with Judaism, 1% with Buddhism, and 2% with an otherwise unspecified spirituality. Participants were also asked to rate how religious they perceived themselves to be on a seven-point Likert scale, from 1 (not at all religious) to 7 (very religious). Responses to this question tended to the extremes, with 38% reporting not being at all religious and 16% being very religious. The remainder of the sample were approximately evenly distributed across the spectrum. Slightly more than half the sample reported their level of religion as being less than 3 on our seven-point scale.

Table 9.4 ranks the factors that participants considered most important in withdrawal-of-treatment decisions. Several factors, such as the presence of consciousness and patient autonomy, have very broad appeal as significant factors in moral decision making. Surprisingly, however, factors such as religion and the sanctity of life—which are often cited in public debates about withdrawal of treatment from such brain-damaged patients—were ranked at the bottom of the list in this regard. It is also interesting that distributive justice was not regarded as important, given the prominence of this principle in contemporary bioethics.

The decreased importance given to distributive justice may also be tied to the general rejection of a "utilitarian" solution to the footbridge dilemma. In

Table 9.4 Factors seen as Important in Decision Making About Withdrawal of Treatment, Ranked From most to Least Important.

Presence of consciousness
Autonomy
Ability to interact with others
Suffering
Dignity
Best interests
Best interests of patient's family
Distributive justice
Religion
Sanctity of life
Longevity

this dilemma, participants are asked whether it is ethical to push a bystander off a bridge into the path of an oncoming vehicle in order to stop the vehicle before it kills five others. In our survey, 77% judged that the protagonist in the dilemma should not sacrifice a bystander to save five other people in the path of a runaway vehicle. This is broadly in line with previous research showing that the majority of respondents reject a utilitarian solution in this dilemma (Greene & Haidt, 2002). A strong positive correlation was found between willingness to push the stranger off the bridge and endorsement of the principle of distributive justice ($p = .003$), considerations of the patient's suffering ($p = .036$), and the patient's quality of life ($p = .037$), although not with any other values. Older respondents were less likely to agree with a utilitarian decision in this case ($p < .05$). Willingness to push in the footbridge dilemma was also correlated with support for withdrawal of treatment in MCS ($p = .002$), LIS ($p = .005$) and total LIS ($p = .05$), but not in VS ($p = .92$).

Although distributive justice needn't be understood in utilitarian terms, the low level of importance ascribed to the principle of distributive justice, as well as repudiation of a utilitarian solution to the footbridge dilemma, suggests that most people reject a cold and calculating cost-benefit analysis as an appropriate consideration in life-and-death decisions. A strict cost-benefit analysis would require the protagonist to sacrifice one person in order to save the maximum number of people. Similarly, a strict cost-benefit analysis would make treatment of disorders of consciousness a low priority on distributive justice grounds.

Rejection of these principles could reflect unease over attempts to place a price on life and unwillingness to let finances dictate who should receive treatment. It could also, however, reflect an inability to take account of the wider societal impacts of the ethical decisions that participants were asked to make. When one is confronted with a single, identifiable patient, there is a sense of immediacy and urgency that is referred to as the "identifiable victim effect" (Genevsky, Vastfjall, Slovic, & Knutson, 2013). In this situation, it would seem callous to refuse treatment by reference to arguments of economy or rationing of medical resources. This sense of immediacy and connection with a particular patient could explain the psychological basis of the generally low priority laypeople give to distributive justice, but it is a departure from the views of a number of experts in the field. For instance, Jennett, one of the first to describe VS, believed that we "certainly should question the utilisation of precious resources [for treatment of disorders of consciousness]" and that we "can no longer afford the luxury of always doing everything for every patient" (Jennett, 1976). Indeed, it might be thought that considerations about the best use of limited resources would be especially paramount in connection with conditions such as VS, where it is doubtful that the patient in question receives any benefit from continuation of treatment.

In further analyses, we investigated possible correlations between views about withdrawal of treatment and participants' religious views. Unsurprisingly, given traditional antagonism toward treatment withdrawal from religious quarters, we found an inverse correlation between religiosity and agreement with treatment withdrawal. This is significant in indicating that opposition to withdrawal of treatment may be based on, or draw upon, a religiously derived system of ethics. Some might argue that the theological foundation of some participants' opposition may reduce the weight that should be given to such opposition in ethical debate. Given the potential for public opinion to shape public policy, religiously motivated moral beliefs may, in some views, be legitimately questioned when they impact secular society.

Table 9.5 shows the correlation between the principles that participants believed important in decision making and their beliefs about the morality of treatment withdrawal. The greater the importance attached to considerations of a patient's suffering, dignity, quality of life, autonomy, best interests, and distributive justice, the greater the level of agreement with treatment withdrawal.

Increasing consideration of the potential for patients in MCS, LIS or total LIS to suffer seems quite compatible with increased agreement with withdrawal of

Table 9.5 Correlations Between Endorsement of Importance of Morally Relevant Factors and Agreement with Withdrawal of Treatment in the Four Conditions.

Factor	Pearson Correlation coefficient	Statistical Significance (2-Tailed)
Positive correlation		
Distributive justice	0.455	<0.001
Patient suffering	0.415	<0.001
Dignity	0.470	<0.001
Quality of life	0.565	<0.001
Autonomy	0.328	<0.001
Best interests	0.236	0.001
Negative correlation		
Religiosity	−0.272	<0.001
Sanctity of life	−0.519	<0.001
No *correlation*		
Best interests of family	0.088	0.219

treatment in such cases. It is perhaps unclear, however, why the suffering of the patient is considered an important factor in decisions surrounding end-of- life care for VS patients, given that these patients were described to participants as lacking any awareness of themselves or their surroundings. This may perhaps be explained by the paradoxical nature of "wakefulness without awareness" in VS, which is difficult to explain, much less grasp, in the context of an Internet-based survey. Alternatively, the lack of potential to suffer harm from treatment withdrawal, which must be a concern in the context of the other disorders, could have been a consideration, and this would be consistent with our results.

Endorsement of the importance of autonomy, best interests, and distributive justice was correlated with greater support for treatment withdrawal at highly statistically significant levels. The four principles (best interests being an amalgamation of beneficence and nonmaleficence) have been formalized into a structure by Beauchamp and Childress (2003), who believed that these principles distill what are common or even universal ethical principles. This model has become the preeminent model of principlism in bioethics and is commonly taught in medical education courses. Given their ubiquity of these principles in medical ethics, it is interesting to find that support for these principles correlates with greater support for treatment withdrawal. One plausible explanation is that acknowledgment of the importance of these principles could reflect familiarity with contemporary bioethics, which itself is often associated with broad acceptance of treatment withdrawal. Alternatively, although somewhat less plausibly, it is possible that the formalization of ethical thought into this structure has framed the debate in such a way as to make support of treatment withdrawal more likely.

In any case, this result demonstrates the different principles and frames of references relied on by those who support and oppose withdrawal of treatment. It is unclear whether the values endorsed by participants inform the ethical decisions they make, or whether they were endorsed as rationalizations for moral beliefs they already held. If the former is true, then these results present a blueprint for how ethicists may influence the ethical and moral beliefs of laypeople. By emphasizing the importance of the principle of distributive justice and considerations of the patient's suffering, dignity, and quality of life—all of which are factors positively correlated with increased acceptance of treatment withdrawal—laypeople may be more likely to agree that this is indeed ethical. By the same token, countering arguments based on religion and religiously derived principles such as the sanctity of life may reduce antagonism toward withdrawal of treatment.

This would bring the attitudes of ordinary people in line with the views espoused by the majority of ethicists.

Correlates of Personality Traits

One novel aspect of the work we carried out was to ask participants to complete a series of widely used personality measures to explore associations between these measures and their opinions on treatment withdrawal. The instruments used included the Big Five Inventory (a general measure of five dimensions underlying personality: openness to experience, conscientiousness, extroversion, agreeableness, and neuroticism); the Interpersonal Reactivity Index (a standard measure of empathy); and the Need for Cognition scale (a measure of the tendency for an individual to engage in and enjoy effortful thinking).

Interestingly, we found that greater "need for cognition" was not generally correlated with greater endorsement of withdrawal of treatment. Nor was it correlated with utilitarian responses to the footbridge dilemma, despite previous work showing a tendency for greater need for cognition to be associated with favoring utilitarian decisions (Bartels, 2008). Our negative association here between engagement in cognitive effort, as measured by the Need For Cognition Scale, and acceptance of treatment withdrawal indirectly suggests that endorsement of this moral position is not driven by greater reflection and that opposition to withdrawal of treatment may not be based merely on immediate gut reactions. However, high need for cognition was negatively correlated with willingness to have treatment withdrawn from oneself in LIS. This suggests those with a high need for cognition may be more able to imagine scenarios in which they are capable of adapting to a life with such a severe disability or, alternatively, to imagine possible cures in the future. Or, it may show that they are more satisfied with a life of contemplation rather than action.

The lack of correlation between scores on the different subscales of the Interpersonal Reactivity Index and support for removal of treatment also indicates that empathy may not play a significant role in these decisions. Many philosophers see empathic concern, particularly, as central to morality (Smith, Macfie, & Raphael, 1976). One would therefore expect that individual differences in empathic concern would make a significant difference to responses to difficult moral questions. Our negative result here suggests that empathy and empathic concern are not significantly involved in the formation of these moral judgements. It is also possible, however, that empathic concern can dispose people to ethical decisions in both directions. That is, greater empathic concern could make it harder to endorse the seemingly harmful act of withdrawing treatment, but it could also magnify one's response to the plight of severely disabled individuals.

It might be expected that traits measured by the Big Five Inventory, given their role in developing personality, would have a stronger effect on developing ethical beliefs. There was a slight positive correlation between a classification as more neurotic or emotionally unstable and support for treatment withdrawal from oneself in the case of VS. This may indicate the role of negative emotions (e.g., stress) in willingness to withdraw treatment. Similarly, the tendency for those who are more agreeable to refuse withdrawal for themselves in VS may implicate the role of positive behaviors such as optimism in refusal.

In summary, the personality traits that we measured did not have strong or consistent associations with moral views about withdrawal of treatment. There were some correlations between agreement with treatment withdrawal in individual conditions and scores on personality measures. These correlations, however, were not strongly statistically significant, nor were they consistent across different situations. Other factors, such as a person's religiosity or the values endorsed as being important in decision making, appear to have a much greater role in forming ethical beliefs. These traits do not seem to be dependent on the different personality types that we tested for. This may suggest that ethical values are not mere artefacts of people's personality traits but instead represent independent factors capable of influencing concrete moral decision making.

LIMITATIONS

This study has several limitations. People may not always have accurate introspective access to the factors that really drive their moral views, and it is difficult to convey complex clinical and ethical notions in a survey form. A move to focus groups may therefore provide an important source of further evidence about lay attitudes to withdrawal of treatment. Another inherent limitation to this sort of cross-sectional study design is the inability to establish a temporal relationship and therefore causality. For this reason, there is no way of knowing whether the values participants considered important in decision making were the foundations of those decisions or simply rationalizations of them.

One issue that seems urgent to clarify is the apparent inconsistencies in opposition to withdrawal of treatment. This is evident in the gap between first- and third-person judgments in these decisions and also in the much higher rate of endorsement of withdrawal of treatment in the *W v. M* case than in MCS generally. Future research should elucidate how people justify these discrepancies, a point that the study reported here cannot fully resolve.

THE NORMATIVE SIGNIFICANCE OF LAY ATTITUDES

The aim of our survey was to investigate the moral attitudes of laypeople to difficult moral questions about withdrawal of treatment from patients with several forms of severe brain damage. We end this chapter with some brief reflections on the potential normative significance of such surveys.

Our survey can be seen as an instance of what is known as "empirical ethics." The foundation of empirical ethics generally, and the use of surveys such as this in particular, is that people's actual moral judgments, beliefs, attitudes, and actions have significance for normative ethics (Salloch, Schildmann, & Vollmann, 2012). This is an axiom that has been challenged by a number of philosophers for committing the meta-ethical fallacy of attempting to bridge the "is/ought" gap. Simply stated, this is the gap between how things are and how they should be. For example, Borry, Schotsmans, & Dierickx (2004) wrote that empirical disciplines "describe how reality is constructed—they describe what 'is.' However, they can never tell how people ought to behave, or what kinds of decisions are morally acceptable." These authors saw empirical research as fundamentally divorced from normative ethics, and for that reason, empirical methodologies are of no direct normative interest.

This criticism stands only if the aim of empirical ethics is to directly translate popular opinion into moral norms. But surveys of the moral attitudes and intuitions of ordinary folk can have a more modest normative aim. To the extent that ethicists assume that the moral intuitions they appeal to are widely shared, surveys of the kind reported here can serve as a "check" on the claims of ethicists. This is not to say that a majority vote against a particular act should constitute a prohibition on it, only that it should require a correspondingly greater weight of philosophical argument to justify it. As Borry et al. (2004) wrote, "Ethics does not always have to simply accept what empirical research produces, but at the same time ethics bears a heavy burden of proof if it proposes a different view."

Furthermore, surveys can tell us which moral views are acceptable, or at least potentially acceptable, to the general population, thus giving us a better sense of whether some moral view (correct or not) is likely to be successfully implemented in policy in the near future, or whether at this point it might be better to promote a weaker moral view that is more acceptable to people. Conversely, surveys can also reveal that some supposedly controversial moral views are actually held by many ordinary people, contrary to what is assumed. This again may have implications for policy. For example, our findings about the low level of support for removal of treatment from patients in MCS and LIS indicates that there would be opposition from the wider population to any public policy supporting such a policy in the near future.

Although lay resistance to withdrawal of treatment in MCS and LIS appears considerable, its strength and nature are open to question. As we saw, support for withdrawal of treatment is significantly increased when withdrawal is considered from the first-person as opposed to third-person perspective, and many more participants endorsed withdrawal of treatment in the concrete *W v. M* case than did so as a general principle in MCS. As well as indicating that this specific verdict is out of step with popular opinion, these results suggest that treatment withdrawal becomes more popular when it is considered in concrete cases rather than in abstract or general ways. It also indicates that people may be more willing to support withdrawal if it can be shown to be in line with a patient's wishes, consistent with our results showing the importance that laypeople generally give to considerations of autonomy. In this way, empirical ethics is able to provide valuable insights into what really concerns ordinary people and how their values can most effectively be engaged.

Our survey also suggests that certain kinds of moral consideration, and certain forms of moral argument, are more likely than others to persuade ordinary people to endorse treatment withdrawal in these clinical cases. Ethicists have put forward a range of arguments to justify treatment withdrawal, which, for a number of reasons, can fail to be convincing to the wider public. These arguments may appeal to factors beyond the scope of the experience of laypeople, or they may employ esoteric concepts that can be misunderstood by those not trained in philosophy. The arguments of ethicists are more likely to influence real-world decisions if they are persuasive not only to professional ethicists but also to the majority of people. Dunn, Sheehan, Hope, & Parker (2012) wrote that the most legitimate aim of empirical ethics is to create arguments that are both "convincing and able to convince." This may take the form of presenting a particular moral viewpoint as one of common sense held by the majority of people. The results we report here about the values and principles that ordinary folk regard as most important to decisions about withdrawal of treatment are of obvious relevance in this connection. For example, considerations relating to autonomy, best interests, and, to a lesser extent, distributive justice seem to have particular weight in acceptance of treatment withdrawal.

Finally, the results of our surveys may contribute to normative inquiry by clarifying the underlying reasons for ethical disagreements about the permissibility of withdrawal of treatment. More specifically, our findings indicate that some of this disagreement is due to the different frameworks of principles relied on. Those who draw on something like Beauchamp and Childress' four principles of bioethics are more likely to agree that treatment withdrawal is morally acceptable. Those who subscribe to a more religiously based ethic, however, are more likely to deny the moral permissibility of such acts.

CONCLUSION

Modern medicine's capacity to artificially extend and sustain lives in conditions of severe brain damage has generated a spirited debate on the morality of doing so. We attempted to contribute to this debate by considering an empirical perspective and, in particular, by trying to clarify the factors that shape the views of laypeople on these moral questions. This study was novel in that it looked at the moral attitudes of laypeople as opposed to healthcare professionals, on whom much of the previous work has focused. It was also novel in its examination of the ethical principles and underlying personality traits that may sustain these ethical beliefs. The views of laypeople are significant in gauging the social acceptability of policies relating to treatment withdrawal and for the development of public policy regulating it.

REFERENCES

Airedale NHS Trust v. Bland. (1993). *All England Law Report, 1*, 821–896.

American Congress of Rehabilitation Medicine. (1995). Recommendations for use of uniform nomenclature pertinent to patients with severe alterations in consciousness. *Archives of Physical Medicine and Rehabilitation, 76*(2), 205–209.

Andrews, K. (1997). Vegetative state: Background and ethics. *Journal of the Royal Society of Medicine of London, 90*(11), 593–596.

Asai, A., Maekawa, M., Akiguchi, I., Fukui, T., Miura, Y., Tanabe, N., & Fukuhara, S. (1999). Survey of Japanese physicians' attitudes towards the care of adult patients in persistent vegetative state. *Journal of Medical Ethics, 25*(4), 302–308.

Ashwal, S., & Cranford, R. (2002). The minimally conscious state in children. *Seminars in Pediatric Neurology. 9*(1), 19–34.

Bartels, D. M. (2008). Principled moral sentiment and the flexibility of moral judgment and decision making. *Cognition, 108*(2), 381–417.

Beauchamp, T. L. (2003). A defense of the common morality. *Kennedy Institute of Ethics Journal, 13*(3), 259–274.

Beauchamp, T. L., & Childress, J. F. (1994). Morality and moral justification. In T. L. Beauchamp and J. F. Childress. *Principles of biomedical ethics*. Oxford: Oxford University Press.

Bekinschtein, T., Leiguarda, R., Armony, J., Owen, A., Carpintiero, S., Niklison, J., . . . & Manes, F. (2004). Emotion processing in the minimally conscious state. *Journal of Neurology, Neurosurgery, and Psychiatry, 75*(5), 788.

Bernat, J. L. (2006). Chronic disorders of consciousness. *Lancet, 367*(9517), 1181–1192.

Boly, M., Faymonville, M. E., Peigneux, P., Lambermont, B., Damas, F., Luxen, A., . . . & Laureys, S. (2005). Cerebral processing of auditory and noxious stimuli in severely brain injured patients: Differences between VS and MCS. *Neuropsychological Rehabilitation, 15*(3-4), 283–289.

Boly, M., Faymonville, M. E., Schnakers, C., Peigneux, P., Lambermont, B., Phillips, C., . . . & Laureys, S. (2008). Perception of pain in the minimally conscious state with PET activation: An observational study. *Lancet Neurology, 7*(11), 1013–1020.

Borry, P., Schotsmans, P., & Dierickx, K. (2004). What is the role of empirical research in bioethical reflection and decision-making? An ethical analysis. *Medicine, Health Care, and Philosophy, 7*(1), 41–53.

Buchanan, A. E., & Brock, D. W. (1989). *Deciding for others: The ethics of surrogate decision making*. Cambridge, UK: Cambridge University Press.

Christakis, N. A., & Asch, D. A. (1995). Physician characteristics associated with decisions to withdraw life support. *American Journal of Public Health, 85*(3), 367–372.

Cruzan v. Director, Missouri Department of Health. (1990). *Wests Supreme Court Report, 10*, 2841–2892.

Demertzi, A., Ledoux, D., Bruno, M. A., Vanhaudenhuyse, A., Gosseries, O., Soddu, A., . . . & Laureys, S. (2011). Attitudes towards end-of-life issues in disorders of consciousness: A European survey. *Journal of Neurology, 258*(6), 1058–1065.

Demertzi, A., Racine, E., Bruno, M. A., Ledoux, D., Gosseries, O., Vanhaudenhuyse, A, . . . & Laureys, S. (2013). Pain perception in disorders of consciousness: Neuroscience, clinical care, and ethics in dialogue. *Neuroethics, 6*(1), 37–50.

Demertzi, A., Schnakers, C., Ledoux, D., Chatelle, C., Bruno, M. A., Vanhaudenhuyse, A., . . . & Laureys, S. (2009). Different beliefs about pain perception in the vegetative and minimally conscious states: A European survey of medical and paramedical professionals. *Progress in Brain Research, 177*, 329–338.

Doble, J. E., Haig, A. J., Anderson, C., & Katz, R. (2003). Impairment, activity, participation, life satisfaction, and survival in persons with locked-in syndrome for over a decade: Follow-up on a previously reported cohort. *Journal of Head Trauma Rehabilitation, 18*(5), 435–444.

Dunn, M., Sheehan, M., Hope, T., & Parker, M. (2012). Toward methodological innovation in empirical ethics research. *Cambridge Quarterly of Healthcare Ethics, 21*(4), 466–480.

Formby, A. P., Cookson, R. A., Halliday, S., Formby, A. P., Cookson, R. A., & Halliday, S. (2015). *Cost analysis of the Legal Declaratory Relief Requirement for Withdrawing Clinically Assisted Nutrition and Hydration (CANH) from Patients in the Permanent Vegetative State (PVS) in England and Wales*. Research Paper 108, Centre for Health Economics, University of York. Heslington, England: University of York.

Gardner: re BWV. (2003). *Victorian Supreme Court*. VSC 173.

Genevsky, A., Vastfjall, D., Slovic, P., & Knutson, B. (2013). Neural underpinnings of the identifiable victim effect: Affect shifts preferences for giving. *Journal of Neuroscience, 33*(43), 17188–17196.

Gipson, J., Kahane, G., & Savulescu, J. (2014). Attitudes of lay people to withdrawal of treatment in brain damaged patients. *Neuroethics, 7*(1), 1–9.

Gold, N., Pulford, B. D., & Colman, A. M. (2014). The outlandish, the realistic, and the real: Contextual manipulation and agent role effects in trolley problems. *Frontiers in Psychology, 5*, 35.

Greene, J., & Haidt, J. (2002). How (and where) does moral judgment work? *Trends in Cognitive Science*, 6(12), 517–523.

Hodges, M. O., Tolle, S. W., Stocking, C., & Cassel, C. K. (1994). Tube feeding: Internists' attitudes regarding ethical obligations. *Archives of Internal Medicine, 154*(9), 1013–1020.

Jennett, B. (1976). Resource allocation for the severely brain damaged [Editorial]. *Archives of Neurology, 33*(9), 595–597.

Jennett, B. (2002). The vegetative state. *Journal of Neurology, Neurosurgery, and Psychiatry, 73*(4), 355–357.

Jennett, B., & Plum, F. (1972). Persistent vegetative state after brain damage: A syndrome in search of a name. *Lancet, 1*(7753), 734–737.

John Paul II. (2004, March 20). *Address of John Paul II to the participants in the International Congress on "Life-Sustaining Ttreatments and Vegetative State: Scientific Advances and Ethical Dilemmas"* [Speech]. Retrieved from http://w2.vatican.va/content/john-paul-ii/en/speeches/2004/march/documents/hf_jp-ii_spe_20040320_congress-fiamc.html

Johnson, L. S. (2011). The right to die in the minimally conscious state. *Journal of Medical Ethics, 37*(3), 175–178.

Kahane, G., & Savulescu, J. (2009). Brain damage and the moral significance of consciousness. *Journal of Medical Philosophy, 34*(1), 6–26.

Kuehlmeyer, K., Borasio, G. D., & Jox, R. J. (2012). How family caregivers' medical and moral assumptions influence decision making for patients in the vegetative state: A qualitative interview study. *Journal of Medical Ethics, 38*(6), 332–337.

Kuehlmeyer, K., Racine, E., Palmour, N., Hoster, E., Borasio, G. D., & Jox, R. J. (2012). Diagnostic and ethical challenges in disorders of consciousness and locked-in syndrome: A survey of German neurologists. *Journal of Neurology, 259*(10), 2076–2089.

Laureys, S., Pellas, F., Van Eeckhout, P., Ghorbel, S., Schnakers, C., Perrin, F., . . . & Goldman, S. (2005). The locked-in syndrome: What is it like to be conscious but paralyzed and voiceless? *Progress in Brain Research, 150*, 495–511.

Lisker, R., Alvarez Del Rio, A, Villa, A. R., & Carnevale, A. (2008). Physician-assisted death: Opinions of a sample of Mexican physicians. *Archives of Medical Research, 39*(4), 452–458.

Lulé, D., Zickler, C., Häcker, S., Bruno, M. A., Demertzi, A., Pellas, F., . . . & Kübler, A. (2009). Life can be worth living in locked-in syndrome. *Progress in Brain Research, 177*, 339–351.

Nelson, L. J., & Cranford, R. E. (1999). Michael Martin and Robert Wendland: Beyond the vegetative state. *Journal of Contemporary Health Law and Policy, 15*(2), 427–453.

Parfit, D. (1987). *Reasons and persons.* [2nd reprint with corrections of the 1984 ed.] Oxford: Clarendon Press.

Payne, K., Taylor, R. M., Stocking, C., & Sachs, G.A. (1996). Physicians' attitudes about the care of patients in the persistent vegetative state: A national survey. *Annals of Internal Medicine, 125*(2), 104–110.

Plum, F., & Posner, J. B. (1972). The diagnosis of stupor and coma. *Contemporary Neurology Series, 10*, 1–286.

Robertson, J. A. (1991). Second thoughts on living wills. *Hastings Center Report, 21*(6), 6–9.

Royal College of Physicians. (1996). The permanent vegetative state. Review by a working group convened by the Royal College of Physicians and endorsed by the Conference of Medical Royal Colleges and their faculties of the United Kingdom. *Journal of the Royal College of Physicians of London, 30*(2), 119–121.

Salloch, S., Schildmann, J., & Vollmann, J. (2012). Empirical research in medical ethics: How conceptual accounts on normative-empirical collaboration may improve research practice. *BMC Medical Ethics, 13*, 5.

Savulescu, J. (1994). Rational desires and the limitation of life-sustaining treatment. *Bioethics, 8*(3), 191–222.

Schroeder, D. (2010). Dignity: One, two, three, four, five, still counting. *Cambridge Quarterly of Healthcare Ethics, 19*(1), 118–125.

Smith, A., Macfie, A. L., & Raphael, D. D. (1976). *The theory of moral sentiments.* Oxford: Clarendon Press.

Sprung, C. L., Cohen, S. L., Sjokvist, P., Baras, M., Bulow, H. H., Hovilehto, S., . . . & Ethicus Study Group. (2003). End-of-life practices in European intensive care units: The Ethicus Study. *JAMA, 290*(6), 790–797.

Stone, J. (2007). Pascal's wager and the persistent vegetative state. *Bioethics, 21*(2), 84–92.

Vincent, J. L. (1999). Foregoing life support in western European intensive care units: The results of an ethical questionnaire. *Critical Care Medicine, 27*(8), 1626–1633.

W v. M. (2011). EWHC 2443 (Fam). In the High Court of Justice of England and Wales, Family Division.

Zeman, A. (2001). Consciousness. *Brain, 124*(Pt 7), 1263–1289.

10

Moral Conflict in the Minimally Conscious State

JOSHUA SHEPHERD

MCS AND CONSCIOUSNESS

After severe traumatic or anoxic brain injury, some patients enter into the minimally conscious state (MCS). MCS patients evince awareness of self and environment by way of inconsistent but discernibly purposeful behaviors—for example, visual fixation and pursuit, command following, and intelligible verbalization (Giacino et al., 2002). According to most, these behaviors are indicative of at least minimal conscious mental life: MCS patients are thought to possess consciousness. Further, by virtue of possessing consciousness, MCS patients are thought to possess a type of moral significance not attributed to patients who lack consciousness (e.g., patients in the vegetative state). How ought this inform decision making regarding MCS patients?

A complication accompanies the fact that the term *consciousness* can be taken to connote a number of distinct phenomena. Most relevant here is Ned Block's (1995) well-known distinction between *access* consciousness and *phenomenal* consciousness. A mental state is access conscious if the information it carries is poised for use in reasoning and in control of behavior. By contrast, a mental state is phenomenally conscious if there is something "it is like" to be in it. Conceivably, these forms of consciousness dissociate in subjects: For all we know, it is possible that an MCS patient can possess access consciousness even though there is nothing "it is like" to be in MCS.

Arguably, both access and phenomenal consciousness are morally significant. According to Walter Sinnott-Armstrong and Franklin Miller (2013), the abilities an agent possesses have moral significance. They argue that killing is morally wrong because it causes the loss of all abilities. Their position could be leveraged to generate a view in which the moral significance possessed by MCS patients is tied (at least in part) to the mental abilities they retain—the abilities that access consciousness enables. Alternatively, Charles Siewert (1998, 2013) has argued that phenomenal consciousness has intrinsic moral importance. According to Siewert, the possession of phenomenal consciousness undergirds a subject's irreplaceability as a person, as well as his or her status as an apt target for empathy.

The moral significance of access consciousness deserves further consideration. In what follows, however, I focus on phenomenal consciousness. I do so in part because a treatment of both aspects in the same paper would become unwieldy. But more importantly, it seems to me that much of the attention given to recent work on MCS—such as the striking finding by Owen et al. (2006; Monti et al., 2010) that some MCS patients have the ability to voluntarily initiate and maintain mental imagery for protracted periods—accompanies the assumption that there is something it is like to be in MCS,[1] and that this is so because phenomenal consciousness is the central conception of consciousness.

In the next section, I sketch a view in which the possession of phenomenal consciousness (henceforth: "consciousness") is necessary for possession of (positive or negative degrees of) subjective well-being. It would seem that the possession of consciousness supplies caregivers reason to enhance the well-being of MCS patients. Unfortunately, as I discuss next, matters are complicated by a certain kind of moral conflict that arises in decision-making situations regarding MCS patient care. In many cases, it seems difficult, and perhaps impossible, to respect an MCS patient's autonomy—as embodied in her autonomously expressed prior wishes or in the wishes she would presently autonomously express were she competent to do so—while promoting the well-being she presently enjoys and will plausibly enjoy in the future. Later, I consider views according to which the moral conflict is only apparent, because considerations of autonomy trump considerations of well-being (or vice-versa). I argue that neither view is satisfying: We are left with genuine moral conflict. However, consideration of these views is salutary, because their weaknesses motivate a mixed view in which considerations of both autonomy and well-being should in many cases be weighed against each other, as well as other relevant moral considerations (e.g., considerations of distributive justice). In the final section, I draw four practical conclusions.

MCS AND THE MORAL SIGNIFICANCE OF CONSCIOUSNESS

In my view, the type of moral significance that MCS patients possess, and VS patients lack, has to do with the potential for subjective well-being that MCS patients possess and VS patients lack. Subjective well-being is what a person has (to some degree) when it is true to say that her life is going well or poorly (to some degree) for her, or from her perspective. In focusing on subjective well-being, I ignore views of well-being that emphasize the importance of the objective properties of a subject's life. Such views are often called objective list theories. As Derek Parfit has it, such theories enumerate the objective goods or bads that enhance or diminish well-being "whether or not these people would want to have the good things, or to avoid the bad things" (Parfit, 1984, p. 499; for a recent defense of an objective list theory, see Rice, 2013). I do not claim that proponents of such views misuse the term *well-being*. I focus on subjective well-being (henceforth: "well-being") because doing so allows me to direct attention to features of the relevant cases that I regard as both morally important and importantly distinct from debates between subjectivists and objectivists about well-being.

What is the connection between consciousness and well-being? In my view, the possession of consciousness is a necessary condition for the possession of some negative or positive amount of well-being: Possession of consciousness *just is* possession of the kind of subjective perspective necessary for possession of well-being to some positive or negative degree. Kahane and Savulescu's observation (2009, p. 13) is apt: "Think of how awry it seems to say 'He led a good life—but there was absolutely nothing it was like to live that life.'" There must be something it is like for a subject in order for things to go well or poorly for that subject.

What determines one's *amount* of well-being? We can distinguish between actual and potential amounts of well-being: Amounts of actual well-being are closely tied to one's actual experiences, and amounts of potential well-being are closely tied to the types of experiences available to one. Consider two mentally and functionally identical MCS patients. By virtue of their identical cognitive, behavioral, and perceptual capacities, there is a sense in which these two patients possess the same amount of potential well-being. But suppose that one is neglected in an uncomfortable bed while the other receives very attentive, state-of-the-art treatment, frequent physical therapy, pain medication, and so on. Thanks to their actual experiences, the actual amount of well-being each enjoys will differ (and, over time, we can expect the amount of potential well-being to shift as well).

When considering the way in which experiences contribute to well-being, it is easy to simplify matters by focusing on simple examples (e.g., intense pleasure and intense pain). Doing so obscures a number of relevant and important issues. For example, are there dimensions other than *positivity of experience* that are relevant here (e.g., diversity of experience)? What is the relationship between a subject's cognitive sophistication and the experience-types available to her? What phenomenal properties determine an experience-type's contribution to well-being? Answering these difficult questions here would take us far afield. Even so, it is worth noting that human beings enjoy a wide range of experience-types, and that this fact is relevant to an understanding of amounts of well-being. Here is a comment by Peter Railton (2009):

> Humans find reward in pleasant experience, to be sure, but also in such things as successful pursuit of abstract ideals, excellence in the exercise of skills and capacities, discovery and the creation of knowledge, friendship, humor, self-expression, aesthetic appreciation, romantic love, and commitment to kith and kin. (p. 94)

Of course, many of the experience-types available to healthy human subjects are not available to MCS patients. But recent neuroscientific work on MCS patients indicates that for at least some in this population, a variety of experience-types remain available. Many MCS patients retain high-level semantic processing of speech (Coleman et al., 2007). At least some MCS patients retain the capacity to deploy top-down visual attention (Monti et al., 2012), as well as the ability to voluntarily initiate and maintain mental imagery for protracted periods for the purposes of answering "yes or no" questions posed by experimenters (e.g., 30 seconds; see Monti et al., 2010; Owen et al., 2006). This seems to indicate a fairly robust mental life—one that includes not only volition, but the higher cognitive capacities that support reasoning (Hampshire et al., 2013). The science of MCS is in its early days, but initial indications support the view that at least some MCS patients retain the capacity to both enjoy substantial amounts of well-being and suffer from a significant deprivation of well-being.

MORAL CONFLICT

We want to know about the moral significance of the consciousness MCS patients possess in part because we want moral guidance regarding the care of MCS patients. But any plausible story about MCS patient care must address not only the moral significance of consciousness but also the interactions between this source of moral significance and other morally relevant considerations (e.g., distributive justice, the value of autonomy). In particular, MCS caregivers must often attend to a conflict that arises between considerations of patient well-being and patient autonomy, where autonomy is understood as a capacity to lead one's own life, often by making decisions (e.g., about care) that define the nature and course of that life (see Dworkin, 1993, p. 222).

The case of *W v. M* (2011) illustrates the difficulties involved. At 43 years of age, M suffered brain damage from viral encephalitis. After she emerged from coma, doctors judged that she had entered the VS. Almost 4 years passed. Her family then sought authorization to remove treatment, at which point further examination of M led to a change in diagnosis, from VS to MCS. The family decided to press on, and the case—the first of its kind in England—went before the English Court of Protection.

Although there was no advance directive in this case, M's prior wishes were clear. Regarding situations similar to her own, M had expressed to family and friends on numerous occasions that she would not want her life to be maintained. However, in court testimony, M's caregivers painted a fairly complex picture of her current well-being. M obviously experienced pain and discomfort, such as when her incontinence pad was changed. When certain songs were played for her, she reliably wept; once M was shown a video of a wedding, and she made a very distressed sound until caregivers turned off the video. But M's caregivers provided evidence of positive experience as well. According to testimony provided to the court, M tapped her wrist in tune with fast music, relaxed when given hand massages, seemed to enjoy being taken into the garden or out into the sun when it was warm, seemed to smile (especially in response to certain familiar male caregivers, certain music, and television programs), seemed to communicate comfort or discomfort based on different types of moans, and seemed to behave in a playful manner with some caregivers.

As stipulated in the Mental Capacity Act 2005 (MCA) (2007), the English Court of Protection was charged with deciding what was in M's "best interests." Although the MCA gives the court some latitude in determining best interests, the MCA's Code of Practice makes clear that the standard centrally involves a patient's past autonomous wishes and values:

> 5.32 . . . [I]n particular, the decision-maker should consider any statements that the person has previously made about their wishes and feelings about life-sustaining treatment. . . .
> 5.41 The person may have held strong views in the past which could have a bearing on the decision now to be made. All reasonable efforts must be made to find out whether the person has expressed views in the past that will shape the decision to be made. This could have been through verbal communication, writing, behaviour or habits, or recorded in any other way (for example, home videos or audiotapes). (2005)

However, in determining M's best interests the court awarded little weight to her prior wishes and values. Instead, the court (*W v. M*, 2011) cited a principle of "preservation of life" (paragraph 7) as relevant, as well as M's (actual and potential) well-being: "I find that she does have some positive experiences and importantly that there is a reasonable prospect that those experiences can be extended by a planned programme of increased stimulation" (paragraph 8). As a result of these considerations, the court refused the application to withdraw artificial nutrition and hydration.

Theorists have sharply criticized the court's neglect of M's prior wishes. According to Alexandra Mullock, "The legal requirement to consider the past wishes and views of incompetent patients should . . . be seen as . . . an obligation to respect autonomy" (2012, p. 2). Emily Jackson agreed and emphasized the need for everyone to establish their wishes in a way the court will be unable to neglect: "The moral of this sad story is that all of us—even when we are completely healthy—need to think about whether we would wish to have [artificial nutrition and hydration] withdrawn if we were ever in a MCS, otherwise we risk being played music that makes us cry and being kept alive in part because stopping moaning when one has just had one's incontinence pads changed is said to be evidence of contentment" (2012, p. 3).

Was the court right in privileging the existence of positive experiences, and the potential for extension of such experiences, over M's autonomous wishes?[2] This depends on one's view about the relative importance of autonomy and well-being in cases involving MCS patient care. As M's case makes vivid, in such cases we come across a moral conflict: As stated earlier, it appears difficult, and perhaps impossible, to respect an MCS patient's autonomy while simultaneously promoting his or her well-being. We can grant that considerations of autonomy and of well-being are both morally relevant. All else being equal, we have reasons to respect a patient's autonomy or to promote a patient's well-being. Our problem is that, in the cases at issue, all is not equal. Considerations of autonomy and well-being seem to conflict. How ought we to adjudicate the conflict?

AUTONOMY TRUMPS WELL-BEING

Consider the following claim.

> *Autonomy trumps well-being.* When an MCS patient's prior wishes about her present state are clear, properly informed, and sufficiently strong, relevant decision makers should heed these wishes to the exclusion of any considerations of well-being.

Three points of clarification are relevant here. First, a patient's prior wishes (that "P" be done) are clear, roughly, when their content is intelligible and sufficiently specific and when there is no conflicting evidence to the effect that the patient wished something incompatible with P. Second, a patient's prior wishes are properly informed when there is sufficient indication that the patient understood the relevant circumstances surrounding her wishes. The patient need not understand every facet of the relevant circumstances, of course. But as a rule of thumb, a patient should be aware of information that might plausibly cause her to reconsider her prior wishes. Third, the requirement that the wishes be sufficiently strong is intended to exclude cases in which (1) the patient expressed the wish that P be done either hesitantly, or tentatively, or in some way that undermines confidence that the wish was genuine or (2) at any time after expressing the relevant wish, the patient expressed hesitance or anything that undermines confidence that the prior wish remained genuine. (In the case of an advance directive, this requirement might be met by conducting periodic reviews of the advance directive with its author.)

In general, medical and legal practice in related cases is consistent with the claim that *Autonomy trumps well-being.* When patients lack competence to make care-related decisions, advance directives are typically given legal and medical authority. The patient's past wishes, as embodied in the advance directive, take precedence over present well-being, even if the patient appears to enjoy a pleasant existence. In the case of *W v. M*, for example, the court noted that if M had made an advance directive indicating a wish to withdraw treatment, the directive would have determined the court's decision. In a discussion of the perils of decision making for both MCS and VS patients, Joseph Fins and Nicholas Schiff gave voice to a common judgment: "It is our strongly held view that if a patient articulated a preference or completed an advance directive before losing decision-making capacity, the prior wishes should guide care" (2010, p. 23).

Even in the absence of an advance directive, many maintain that a patient's prior wishes (insofar as they can be determined) should trump considerations of well-being. In the United States, when patients lack both an advance directive and the ability to make care-related decisions, it is common to utilize a "substituted judgment" standard. According to this standard, caregivers determine what to do by reference to what the patient would have decided had she been able to make the relevant decision. This standard clearly prizes autonomy over well-being: The patient's hypothetical autonomous decision is taken to be of primary moral importance.[3]

Why should we think, as many seem to, that the notion, *Autonomy trumps well-being*, is correct? Work on care-related decision making in similar cases proves useful here. For example, although severe dementia differs in many important ways from MCS, structural similarities between severe dementia and MCS are illuminating. Consider Ronald Dworkin's (1993) autonomy-friendly view regarding decision making for incompetent, severely demented patients.

Crucial to Dworkin's position is a distinction between a subject's critical interests and her experiential interests. The latter are tied to experiences we find enjoyable as experiences (e.g., watching football, eating well). The former are tied to "convictions about what makes a life good on the whole"—and these interests "represent critical judgments rather than just experiential preferences" (Dworkin, 1993, pp. 201–202). Dworkin has us consider the case of Margo, a severely demented patient whose prior critical interests not to live on in such a state conflict with her present experiential interests. At present, Margo is happy. Even so, Dworkin maintains that Margo's previously held critical interests take moral precedence. "If I decide, when I am competent, that it would be best for me not to remain alive in a seriously and permanently demented state, then a fiduciary could contradict me only by exercising an unacceptable form of moral paternalism" (p. 231).

Three types of consideration seem to ground Dworkin's judgment. First is a view of the moral importance of autonomy, which for Dworkin has to do with an agent's capacity to express her character in leading a life. "Recognizing an individual right of autonomy . . . allows us to lead our own lives rather than be led along by them, so that each of us can be . . . what we have made of ourselves" (Dworkin, 1993, p. 224). Second is a view of what makes a life go better or worse. Whereas Dworkin thinks that recognizing and satisfying the critical interests we ought to have genuinely makes our lives go better, the same is not true of experiential interests. According to Dworkin, "My life is not a worse life to have lived—I have nothing to regret, still less to take shame in—because I have suffered in the dentist's chair" (p. 201). Third is a view of the moral importance of a life considered *as a whole*. For Dworkin, a life has moral value above and beyond the value that might attach to any of the events in it.

Because experiential interests are morally of little import, satisfying them is of little import as well. More important are a patient's previously held critical interests. The satisfaction of these interests can make the patient's life, when considered as a whole, go better. Thus, Dworkin rejects the view that "in the circumstances of dementia, critical interests become less important and experiential interests more so, so that fiduciaries may rightly ignore the former and concentrate on the latter" (p. 232). Interestingly, he rejects the relative moral significance of experiential interests by way of an analogy between patients in VS and severely demented patients: "Persistently vegetative patients have no sense of their own critical interests, but that is not a good reason for ignoring their fate, and it is not a good reason for ignoring the demented, either" (p. 232).

Insofar as Dworkin rejects the thought that the presence of consciousness makes a morally significant difference, of course, he appears to stray very far from commonsense morality.[4] Moreover, his justification for doing so appears, in my view, rather thin. One reason is that I fail to find claims about the good of a subject's life as a whole compelling.[5] Making a full case for this claim is beyond the present scope, but consider briefly the fact that a single life often has many phases and takes many shapes. Values and religious commitments change, as do careers, relationships, capacities, and so on. Short of a well worked-out theory of the best possible shapes for a life, it is unclear to me what grounds judgments about the good of a subject's life as a whole: The worry is that the tacit normative theory of life's shape informing these judgments will not withstand critical scrutiny.

But grant momentarily that something like the good of a subject's whole life exists. It remains possible to reject Dworkin's judgments about what that implies. We can presume for present purposes that the patient in question experiences some positive level of well-being: Positive experiences are available to her, and we know how to provide them. There is thus an obvious sense in which the positive well-being she has experienced over the course of her life continues to rise. Perhaps her whole life is better if it contains several years of pleasant, even if cognitively impaired, experience.

Dworkin will not like this suggestion, of course. Because Dworkin finds experiential interests of minor moral importance, he accords them little weight. In my view, this part of Dworkin's approach is implausible: The satisfaction of experiential interests is centrally important to a subject's well-being. Human agents go to great lengths—justifiably, it seems to me—to fulfill experiential interests (e.g., tasting a rare scotch, skiing a difficult backcountry chute). And it is arguable that the satisfaction of a critical interest is valuable *in part* because of the experiences such satisfaction engenders—experiences of meaningfulness, achievement, love, and so on.[6] Notice again, however, that one can agree with Dworkin about the minor importance of experiential interests while disagreeing with his judgment about whether considerations of autonomy trump those of well-being. Whatever the moral significance of our critical interests, we need additional reasons to think this significance persists across massive changes to a subject such as occur when patients enter MCS. And we might also wonder—even granting that a whole life has its own kind of moral value—why the value of a whole life is so much more important than the value of a subject's present experiences. Perhaps it is not.

In this connection, consider a case in which an MCS patient's prior wishes were to remain alive, no matter what. And suppose that the patient's case is a horrible one. She is in constant pain. Following Dworkin, we should keep her alive out of respect for her autonomy: Doing so will give her whole life the shape she intended it to have. In such a case, it is difficult to disagree with Seana Shiffrin's verdict: "It seems cruel in such cases to force such people to live through agony so that they will fulfill a critical interest . . . they no longer recognize, accept or even understand" (Shiffrin, 2004, p. 210). In my view, this kind of case demonstrates that it is not in general true that considerations of autonomy *trump* considerations of well-being. The proposition that *Autonomy trumps well-being* is false.[7]

One might object to this line of reasoning as follows.[8] A properly informed subject who decides 'to remain alive, no matter what' has expressed a wish to endure constant pain rather than having treatment withdrawn. But this is irrational: All else being equal, it is irrational to choose pain over pain's cessation. And irrational decisions are not properly expressive of autonomy.

It is true that autonomy requires the capacity for rational behavior *in general* (see Berofsky 1995, p. 10). But particular decisions may properly express autonomy even if they fall short of optimal rationality. Smoking is an irrational activity, indicative of irrational desires, preferences, and decisions, and yet decisions to smoke can nonetheless express an agent's autonomy. Because rationality and autonomy come apart, it seems possible that a subject can irrationally, but autonomously, decide to remain alive in the face of great pain.[9] And if so, then cases likely exist in which considerations of well-being override considerations of autonomy.

WELL-BEING TRUMPS AUTONOMY

Although *Autonomy trumps well-being* is probably the majority view (at least among medical and legal professionals), one finds pockets of dissent.[10] In this section I consider an argument for a view on the opposite end of the spectrum.

> *Well-being trumps autonomy.* When an MCS patient's present and potential future well-being is clear, relevant decision makers should aim to promote patient well-being to the exclusion of any considerations of autonomy.

The line of argumentation I am interested in (although others exist[11]) goes as follows.[12]

Not Really Autonomous

1. An agent J's prior wishes regarding her care have moral weight regarding decisions about her care only if they are rightly considered expressions of J's autonomous judgments or decisions.

2. An agent J's judgment or decision is rightly considered autonomous only if J's judgment or decision is appropriately informed.
3. For any human agent J, no judgment or decision J makes regarding J's care in MCS can be appropriately informed.
4. Thus, for any MCS patient J, J's prior wishes regarding care do not have moral weight for decisions regarding her care.

Should we accept this argument? I find point (1) plausible: Non-autonomous judgments or decisions are not the kinds of things typically taken to be enforceable on a patient's behalf. I find point (2) plausible as well. Although some work is needed to specify what counts as being appropriately informed, some such condition seems to apply. Judgments or decisions that proceed from bad information are paradigmatically nonautonomous. Much of the work in this argument is done, it seems, by point (3). Why think that no human agent—even one apprised of the relevant scientific and clinical literature—can meet the relevant information condition regarding her own care while minimally conscious?

Begin by appreciating our ignorance about what it is like to be minimally conscious. MCS patients have undergone radical cognitive, perceptual, and behavioral changes, and the nature of their day-to-day experience is difficult to fathom. It is unclear whether their experience is analogous to that of less cognitively sophisticated creatures, or of other brain-damaged humans, or whether analogies simply fail here. The patient M reliably wept when she heard Elvis sing, "You were always on my mind." What was hearing this song like for her? Was she experiencing vivid memories of some painful experience? Was this simply a reaction to stimuli, the result of the song's tune triggering some behavioral schema oddly unconstrained because of her injury? Might it have been pleasant for her to weep—a kind of rudimentary experience of catharsis? At present, it is impossible to say.

How much should we make of our ignorance concerning what it is like? Matters are complicated by our present lack of a rigorous account of what level or quality of information is sufficient to render a judgment, or a decision based upon it, autonomous. Even so, it is plausible to think that many judgments or decisions about care when in MCS that people *would and will in the near future make* would and will be insufficiently informed, simply because they would and will be based on either misinformation about MCS or untutored (and implausibly vivid) acts of imagination.

In a recent study of lay attitudes toward withdrawal of treatment decisions, Jacob Gipson, Guy Kahane, and Julian Savulescu (2013) gave participants a clinical description of MCS and asked them to respond to a statement to the effect that they would want treatment withdrawn if they were in such a state.

Some 41% agreed, 36% were unsure, and 22% disagreed. Participants were also given a clinical description of locked-in syndrome (LIS), a condition in which patients retain normal consciousness and cognitive functioning but are almost totally paralyzed. Often, such patients can communicate with others only by moving their eyes. Interestingly, participants' responses to a statement that they would want treatment withdrawn if they were in such a state mirrored their responses to the statement about MCS: 36% agreed, 39% were unsure, and 25% disagreed.

We know—although presumably most of the participants did not—that patients in LIS report relatively high degrees of subjective well-being. Reporting on a recent survey of LIS patients, Bruno et al. (2011a) noted: "Our data show that a non-negligible group of chronic LIS survivors self-report a meaningful life and their demands for euthanasia are surprisingly infrequent" (p. 7). More specifically, 72% of their LIS patients reported positive levels of happiness, 8% reported having suicidal thoughts often (and 24% reported suicidal thoughts occasionally), and 13% reported feeling depressed. It is thus plausible that a high proportion of the participants in Gipson et al.'s study who judged that they would want treatment withdrawn if in LIS would not, in fact, want treatment withdrawn when in LIS.[13] This supports the following two claims. First, the judgments these participants made regarding LIS were insufficiently informed regarding what it is like to be in LIS and thus nonautonomous in the relevant sense. Second, the judgments these participants made regarding MCS, because they were based on a similar lack of relevant information concerning what it is like to be in MCS, are insufficiently informed and thus nonautonomous in the relevant sense.

The problem is compounded by medical and legal experts who offer insufficiently informed assertions from a position of authority. Consider Emily Jackson's assertion, made in the context of criticizing the court's decision in *W v M*: "Imagining myself in M's shoes, I would regard a life in which I was totally dependent on others for all aspects of daily care; immobile; doubly incontinent; moved by a hoist; being played songs that made me cry and uttering occasional words like 'where am I' and 'bloody hell' as, to put it bluntly, a living hell" (Jackson, 2012, p. 1). Jackson's assertion appears to be based primarily on an untutored act of imagination, with additional justification offered by a selective portrayal of M's condition.

In a frequently quoted passage, Ashwal and Cranford (2002) asserted that "if there were a better understanding of MCS, especially the critical issues of consciousness and likelihood for pain and suffering, a broader consensus would develop, that being in a permanent MCS would actually be worse than being in a permanent VS" (p. 29). Aside from an odd claim about what we would judge if we knew more, Ashwal and Cranford's assertion appears to be based on the thought that because they are conscious, MCS patients can feel pain. But MCS patients can feel pleasure as well. Plausibly, some can undergo a wider range of experience-types than pain and pleasure. Ashwal and Cranford's assertion is unhelpful.

It seems plausible, then, that some (perhaps even most) judgments or decisions about care for MCS patients are insufficiently informed. (This raises important practical questions about how best to inform authors of advance directives as well as MCS patient caregivers and family: see the next section of this chapter for discussion.) But this is not enough to deliver the conclusion of *Not Really Autonomous*. One might plausibly maintain that our inability to know what it is like to be minimally conscious is an epistemic constraint future science will overcome. Once we know more about the brain, more about the etiology of brain injury, more about prospects for recovery, more about methods of treatment, more about the cognitive abilities of MCS patients, we will be able to make rough-and-ready judgments about what it is like. Already in M's case, caregivers estimated that 30% of M's days were unpleasant. Perhaps soon we will be able to say "MCS patient Z has cognitive abilities A, B, and C, but lacks D and E; she is in pain for X minutes a day, principally when she undergoes experience-types F and G; she enjoys experiencing H and I, and there is reason to expect that with proper treatment she will one day be able to spend much of her time enjoying L, M, and N." If this is right, then it would seem *Not Really Autonomous* fails. In principle, it should be possible to make judgments or decisions about MCS care that meet plausible criteria for being appropriately informed.

A proponent of *Not Really Autonomous* might reply that the change to experience brought on by brain injury is so radical that our characterizations of it—no matter how empirically informed—will fail to meet plausible criteria for being appropriately informed. The thought here is that whatever characterizations of life in MCS we are able to give will fail to respect the radically different phenomenal character of life in MCS. If we cannot know what it is like to be in MCS—what it is like for an MCS patient to be in pain; to experience pleasure; to experience auditory, tactile, or visual stimulation; to have one's mind wander; or to focus one's attention on something—then we cannot make an informed decision about whether we want to carry on in such a condition.[14]

But there are two significant problems with this reply. First, if it is taken seriously, it threatens to undermine the argument's chief aim. For, if we face radical ignorance about what it is like to be minimally conscious, then we face radical ignorance about how to enhance the well-being of an MCS patient. But in the present context, the point of *Not Really Autonomous* is to undermine considerations of autonomy without simultaneously undermining considerations of well-being.

Second, it is dubious that we can know *nothing* about what it is like for an MCS patient. Certainly our ignorance is vast, but we have little reason to believe that experience in a human being would take a form totally inaccessible to us. We believe, after all, that the conscious lives of healthy adults share many similarities in spite of huge differences in cognitive, perceptual, and behavioral capacities. Plausibly, we should believe the same thing about MCS patients. We should expect that pain is bad for them, that pleasure is good, that a variety of experiences is better than a life of monotony, that greater cognitive sophistication tracks greater potential for well-being, and that behavior—even if confined primarily to neural responses to stimuli—is a fair indicator of what things are like for such patients.

Not Really Autonomous fails to secure victory for the position represented by *Well-being trumps autonomy*. Even so, a weaker form of the argument—one that emphasizes the epistemic difficulties we face when making decisions about MCS care, without claiming that they are in principle surmountable—is very plausible. Decisions about MCS care that wish to accord significance to a patient's prior wishes should pay far closer attention than is now common to the information on which those wishes were based.

CONCLUSION

In the view I have sketched, a certain amount of well-being is available to MCS patients by virtue of their possession of consciousness. This fact generates a moral reason to promote MCS patient well-being. But I have also noted the difficulties that arise in decision-making situations concerning MCS patient care. Because considerations of autonomy often conflict with considerations of well-being, we are faced with difficult choices. It would be easier if one type of consideration systematically outweighed another. However, troubles beset both the view that autonomy trumps well-being and the view that well-being trumps autonomy. The failures of these extreme views motivate a mixed view, in which considerations of both autonomy and well-being should in many cases be weighed against each other, as well as other relevant moral considerations (e.g., considerations of distributive justice). To finish, I wish to draw four practical conclusions from this discussion.

First, more attention should be paid to the informational deficits facing authors of advance directives and MCS patient caregivers and family. Given the gravity of decisions about MCS patient care, such attention might focus on (1) better ways to educate the relevant decision makers about emerging empirical work on MCS (e.g., work of the sort featured in this volume) and (2) the implementation of practical decision procedures that are sensitive to the informational needs of the relevant decision makers. Short of the proper educational and decision-making procedures, injunctions to consider "what the patient would have wanted" or what is in the patient's "best interests" might not be good enough. It is, of course, difficult to say from the armchair what decision-making procedures would be best. In my view, such procedures should be developed by those who are sensitive to the very practical constraints and difficulties attending these decisions. For example, Joseph Fins (2006) noted a difficulty arising from local features of health care in the United States:

> [B]ecause of the geographic separation of acute care and rehabilitation settings, many acute care clinicians have little idea about the course of their patients after hospital discharge. This can lead to distortions among acute care practitioners about what might be achieved over time. This may breed a sense of nihilism about the value of ongoing care because patient prognosis is based upon their limited perspective and contextual experiences. (p. 174)

Decision-making procedures should be designed to reflect the fact that informational deficits (such as overreliance on a limited clinical perspective) often influence care decisions in important ways. Ideally and minimally, regarding decisions to withdraw treatment, both physicians and family members (and, if competent, MCS patients themselves) should be given the time and opportunity to assess the same body of relevant information and to jointly consider diagnostic prospects.

Second, the moral authority generally accorded to considerations of autonomy in end-of-life decision making is, in cases of MCS patient care, out of place. Short of good reasons to think so, we should not let advance directives or clearly expressed prior wishes trump considerations of well-being. Rather, advance directives or clearly expressed prior wishes should be assessed based on the quality of the information on which they were based. Further, even when prior wishes are thought to meet the relevant informational requirement, these wishes should not trump considerations of well-being. Prior wishes should be allowed to offer guidance without dictating the decision to be made.

Third, in such decision-making situations, more attention should be given to considerations of distributive justice. I have not focused on such considerations here, but they are clearly relevant. Any decision to withdraw treatment from an MCS patient who enjoys some amount of positive well-being harms that patient to some degree—at least in the sense that it causes the patient loss of potential well-being. Some argue that considerations of autonomy, or the shape of a patient's whole life, justify this harm. Considerations of distributive justice are another potential justifier. Dominic Wilkinson and Julian Savulescu (2012) argue that in some cases—and they suggest M's case is one of these—considerations of distributive justice support the withdrawal of treatment. The reasoning is straightforward. Caring for an MCS patient is expensive. Funds are limited. If all else is equal, we should distribute funds in a fair way: We must answer the "relative question of whether one life is more worth living or less expensive to support than another" (p. 2). Of course, whether one life is more worth living than another depends crucially on the amount of well-being available to a subject, as well as on how the distribution of funds will influence relevant subjects. More work is required to sort out the best ways to think about how distributive justice should influence care-related decisions for MCS patients.

Fourth, the moral importance of a patient's amount of well-being provides an urgent moral reason to support research that improves our diagnostic capacities. MCS is not a static condition. Retained cognitive, perceptual, and behavioral capacities vary widely among MCS patients. Recognizing this, Bruno et al. (2011b) recently proposed a refinement of the diagnosis of MCS, into categories of MCS+ and MCS−. They draw the distinction as follows:

> MCS+ was defined by the presence of (a) command following, (b) intelligible verbalization or (c) gestural or verbal yes/no responses. In contrast, MCS− patients only show minimal levels of behavioural interaction characterized by the presence of non-reflex movements such as: (a) orientation of noxious stimuli, (b) pursuit eye movements that occur in direct response to moving or salient stimuli, (c) movements or affective behaviors that occur appropriately in relation to relevant environmental stimuli. (p. 1375)

The categorization of MCS is likely to undergo additional refinement as our understanding of MCS increases. Plausibly, token cases of MCS permit large differences in potential for well-being. Therefore, although MCS is rightly thought to be morally different from VS, it is possible that token cases of MCS admit of moral differences at least as large. Decision-making regarding MCS patient care should be based on as accurate an understanding as possible of the capacities each MCS patient retains, as well as those capacities they might, with adequate treatment, one day recover.

NOTES

1. Tim Bayne (2013) has argued that signs of agency (e.g., command following, intelligible verbalization) are markers of phenomenal consciousness. So, whatever the normative differences between access and phenomenal consciousness, if Bayne is right, then many of the reasons we have to attribute access consciousness to a subject will also serve as reasons to attribute phenomenal consciousness.
2. A further question is whether the court was legally right in citing a principle of preservation of life. Because I am interested in certain moral rather than legal

aspects of this case, I do not consider this question. For a discussion of some of the case's legal ramifications, see Sheather (2012).

3. For criticism of the substituted judgment standard, see Torke, Alexander, and Lantos (2008).
4. A recent survey by Gipson, Savulescu, and Kahane (2013) found that 40.2% of participants found it morally acceptable to remove treatment from VS patients, but only 20.6% found it morally acceptable to remove treatment from MCS patients. Further, 17.6% found it morally unacceptable to remove treatment from VS patients, and 41.2% found it morally unacceptable to remove treatment from MCS patients. This constitutes some evidence for the claim that the presence of consciousness has a significant role in commonsense morality.
5. Although differing from Dworkin at many places, Jeff McMahan agrees with his verdict concerning the relevant cases. Although McMahan judges that what is good for a relatively happy demented patient at present is the continuance of life, he maintains that the patient's life as a whole, as well as the part of her life that occurred before dementia, have moral value. And he judges that the goodness of both of these are negatively affected by a failure to implement the patient's prior wishes. Further, McMahan judges that because the healthy part of the patient's life "is overwhelmingly the dominant part, its good should have priority . . . the Demented Patient's present good ought to be sacrificed for the greater good of her earlier self, which is also the greater good of her life as a whole" (McMahan, 2002, pp. 502–503). Insofar as McMahan's judgment depends on a view of the good of life as a whole, my criticisms of Dworkin apply.
6. Recall Wilfred Sellars' reaction to Daniel Dennett's eliminativism about qualia (the purportedly intrinsic, ineffable properties of experience): "But Dan, qualia are what make life worth living!" Dennett commented: "If you didn't have qualia, you would have nothing to *enjoy* (but also no suffering, presumably). It is generally supposed—though seldom if ever expressed—that it would not be any fun to be a zombie" (Dennett, 2005, p. 91).
7. One might worry that because of moral asymmetries between negative and positive hedonic experiences, this case does not generalize to cases involving positive amounts of well-being. Perhaps, for example, there is a duty to prevent negative hedonic experiences if possible, but no duty to promote positive hedonic experiences if possible. If so, perhaps an MCS patient's positive well-being never overrides considerations of autonomy. But I think this worry, as expressed, goes too far. Plausibly a patient's prior wishes are more easily outweighed when that patient is in pain, but it is too strong (in my view) to claim that no amount of positive well-being could outweigh a patient's prior wishes.
8. Thanks to Walter Sinnott-Armstrong for suggesting this line of response.
9. Both irrationality and the pain a patient endures come in degrees. So one might argue that, at a certain point, the irrationality of a decision undermines autonomy: Some decisions are too irrational to count as autonomous. And perhaps decisions to endure a great amount of pain cross the relevant threshold. But all I need is a case that does not cross this threshold, and in which the patient's well-being overrides the (somewhat) irrational, autonomous decision to "stay alive, no matter what." It is plausible that some cases fall at this point along the spectrum.

10. Consider Rebecca Dresser's complaint: "Legal decision-makers have been preoccupied with safeguarding incompetent patients' rights of self-determination and privacy, largely overlooking these patients' more immediate interests in having their present well-being maintained. This legal approach is . . . insufficiently protective of the incompetent patient's genuine interests" (Dresser, 1986, p. 373).
11. Some suggest that considerations of autonomy are either irrelevant or diminished because, in cases of severe brain injury, the patient is no longer the same person as she was when she expressed her prior wishes. For discussion of this kind of thought, see Dresser (1986), DeGrazia (1999), Shiffrin (2004).
12. For an argument similar to *Not Really Autonomous* concerning severely demented patients, see Wrigley (2007).
13. Of course, someone in LIS could rationally demand the withdrawal of treatment. Moreover, we might be morally bound to honor such a decision.
14. In a recent paper, L. A. Paul (2015) emphasized our ignorance about "what it will be like" for a whole class of transformative experiences (she focused on having a child). According to Paul, in light of our ignorance, decisions about courses of action that involve transformative experiences are neither rational nor irrational. There are obvious affinities between this argument and the one explored in the previous paragraph, but the point made there is distinct from Paul's. Autonomy and rationality are distinct properties of a judgment or decision.

REFERENCES

Ashwal, S., & Cranford, R. (2002). The minimally conscious state in children. *Seminars in Pediatric Neurology, 9*(1), 19–34.

Bayne, T. (2013). Agency as a marker of consciousness. In A. Clark, J. Kiverstein, & T. Vierkant (Eds.). *Decomposing the will* (pp. 160–182). Oxford: Oxford University Press.

Berofsky, B. (1995). *Liberation from self: A theory of personal autonomy.* Cambridge: Cambridge University Press.

Block, N. (1995). On a confusion about a function of consciousness. *Behavioral and Brain Sciences, 18,* 227–287.

Bruno, M.-A., Bernheim, J. L., Ledoux, D., Pellas, F., Demertzi, A. & Laureys, S. (2011a). A survey on self-assessed well-being in a cohort of chronic locked-in syndrome patients: Happy majority, miserable minority. *BMJ Open, 1,* doi: 10.1136/bmjopen-2010-000039

Bruno, M.-A., Vanhaudenhuyse, A., Thibaut, A., Moonon, G., & Laureys, S. (2011b). From unresponsive wakefulness to minimally conscious PLUS and functional locked-in syndromes: Recent advances in our understanding of disorders of consciousness. *Journal of Neurology, 258*(7), 1373–1384.

Coleman, M. R., Rodd, J. M., Davis, M. H., Johnsrude, I. S., Menon, D. K., Pickard, J. D. & Owen, A. M. (2007). Do vegetative patients retain aspects of language comprehension? Evidence from fMRI. *Brain, 130*(10), 2494–2507.

DeGrazia, D. (1999). Advance directives, dementia, and "the someone else problem." *Bioethics, 13*(5), 373–391.

Dennett, D. (2005). *Sweet dreams: Philosophical obstacles to a science of consciousness*. Cambridge, MA: MIT Press.

Dresser, R. (1986). Life, death, and incompetent patients: Conceptual infirmities and hidden values in the law. *Arizona Law Review, 28*, 373–405.

Dworkin, R. (1993). *Life's dominion: An argument about abortion, euthanasia, and individual freedom*. New York: Knopf.

Fins, J. J. (2006). Affirming the right to care, preserving the right to die: Disorders of consciousness and neuroethics after Schiavo. *Palliative and Supportive Care, 4*(2), 169–178.

Fins, J. J., & Shiff, N. D. (2010). In the blink of the mind's eye. *Hastings Center Report, 40*(3), 21–23.

Giacino, J., Ashwal, S., Childs, N., Cranford, R., Jennett, B., Katz, D. I., . . . & Zasler, N. D. (2002). The minimally conscious state: Definition and diagnostic criteria. *Neurology, 58*(3), 349–353.

Gipson, J., Kahane, G., & Savulescu, J. (2014). Attitudes of lay people to withdrawal of treatment in brain damaged patients. *Neuroethics, 7*(1), 1-9.

Hampshire, A., Parkin, B. L., Cusack, R., Espejo, D. F., Allanson, J., Kamau, E., . . . & Owen, A. M. (2013). Assessing residual reasoning ability in overtly non-communicative patients using fMRI. *NeuroImage, 2*, 174–183.

Jackson, E. (2012). The minimally conscious state and treatment withdrawal: *W v M*. *Journal of Medical Ethics, 39*, 559–561.

Kahane, G., & Savulescu, J. (2009). Brain-damaged patients and the moral significance of consciousness. *The Journal of Medicine and Philosophy, 33*, 1–21.

McMahan, J. (2002). *The ethics of killing: Problems at the margins of life*. Oxford: Oxford University Press.

Mental Capacity Act 2005 (2007). Code of practice. Norwich: Stationary Office.

Monti, M. M., Pickard, J. D., & Owen, A. M. (2012). Visual cognition in disorders of consciousness: From V1 to top-down attention. *Human Brain Mapping, 34*, 1245-1253.

Monti, M. M., Vanhaudenhuyse, A., Coleman, M. R., Boly, M., Pickard, J., Tshibanda, J.-F., . . . & Laureys, S. (2010). Willful modulation of brain activity in disorders of consciousness. *New England Journal of Medicine, 362*, 579–589.

Mullock, A. (2012). Best interests and the sanctity of life after *W v M*. *Journal of Medical Ethics, 39*, 553-554.

Owen, A. M., Coleman, M. R., Boly, M., Davis, M. H., Laureys, S., & Pickard, J. D. (2006). Detecting awareness in the vegetative state. *Science, 313*, 1402.

Parfit, D. (1984). *Reasons and persons*. Oxford: Oxford University Press.

Paul, L. A. (2015). What you can't expect when you're expecting. *Res Philosophica, 92*(2). doi: http://dx.doi.org/10.11612/resphil.2015.92.2.1

Railton, P. (2009). Practical competence and fluent agency. In D. Sobel & S. Wall (Eds.). *Reasons for action* (pp. 81–115). Cambridge: Cambridge University Press.

Rice, C. M. (2013). Defending the objective list theory of well-being. *Ratio, 26*(2), 196–211.

Sheather, J. C. (2012). Withdrawing and withholding artificial nutrition from patients in a minimally conscious state: Re: M and its repercussions. *Journal of Medical Ethics, 39*, 543-546.

Shiffrin, S. (2004). Autonomy, beneficence, and the permanently demented. In J. Burley (Ed.). *Ronald Dworkin and his critics* (pp. 195–217). Oxford: Blackwell Publishers.

Siewert, C. (1998). *The significance of consciousness*. Princeton, NJ: Princeton University Press.

Siewert, C. (2013). Speaking up for consciousness. In U. Kriegel (Ed.). *Current controversies in philosophy of mind*. New York: Routledge.

Sinnott-Armstrong, W., & Miller, F. G. (2013). What makes killing wrong? *Journal of Medical Ethics, 39*, 3–7.

Torke, A. M., Alexander, G. C., & Lantos, J. (2008). Substituted judgment: The limitations of autonomy in surrogate decision making. *Journal of General Internal Medicine, 23*(9), 1514–1517.

W v. M and Others. (2011). EWHC 2443 (Fam). In the High Court of Justice of England and Wales, Family Division.

Wilkinson, D., & Savulescu, J. (2012). Is it better to be minimally conscious than vegetative? *Journal of Medical Ethics, 39*, 557-558.

Wrigley, A. (2007). Personal identity, autonomy and advance statements. *Journal of Applied Philosophy, 24*(4), 381–396.

11

What Is Good *for Them*? Best Interests and Severe Disorders of Consciousness

JENNIFER HAWKINS

Most people have very little knowledge of what neurologists refer to as "disorders of consciousness" and, consequently, little grasp of what it would mean to have such a disorder. Perhaps for that reason, insufficient attention has been given to questions about the best interests of such patients—about what would be *best for them*. This essay aims to rectify that omission. Attention to this question is all the more important in light of recent research that has called into question much of the previous understanding of what the lives of such patients may be like.

Of course, in the United States and in many other parts of the developed world, it is assumed that decision making for incompetent patients (who were previously competent) should, as much as possible, aim to reflect what the patient himself either wanted or would have wanted if he had thought it through. The preferred tools for decision making are advance directives and surrogates tasked with informing physicians what the patient would have wanted. In such a system, it is sometimes assumed that we do not need general philosophical reflection about best interests, because individuals make decisions for themselves, and caregivers simply follow these directions. However, this assumption is shortsighted. For even in this system, careful reflection about best interests has several important roles to play.

First, when individuals write advance directives, they try to understand what would really be in their own interests should various things happen to them. Later, at the bedside, caregivers may simply want to know what the patient wanted. But the patient, in formulating his directive is not asking himself what he wants; rather, he is trying to understand what he *should* want. He wants to know what really would be in his best interests. Second, many individuals tell their loved ones—the ones who eventually become their surrogates—that if the need ever arises, the surrogate is to make the decision that he or she thinks would be *best* for the patient. Here again, although caregivers at the bedside may simply want the surrogate to do what the patient wanted, in many cases what the patient wanted was for the surrogate to make a judgment about best interests. Finally, there are always some cases where no advance directive exists and no one is very clear about what the patient wanted or would have wanted. In such cases, there is no alternative but to try to assess the patient's current best interests. For all of these reasons, we must reflect on these matters carefully. That way, when we as individuals are faced with such decisions, either for ourselves prospectively or for those who have entrusted themselves to our care, we will have something considered to guide us.

Most educated people are at least vaguely aware of one particular disorder of consciousness, the permanent vegetative state (PVS),[1] if only because of the media attention given in the past to cases such as those of Karen Ann Quinlan, Nancy Cruzan, and Terri Schiavo (Pence, 2008, Chapter 2). As popularly understood, PVS is the complete, permanent loss of all capacity for conscious awareness. Some patients end up in PVS as a result of losing oxygen to the brain for a prolonged period (as was true in the three cases mentioned), whereas others may enter PVS as a result of traumatic brain injury.[2] Unlike coma patients, PVS patients do have wake-sleep cycles as well as certain other reflexes and involuntary movements. But the general assumption until quite recently was that such patients lacked any degree of awareness.

The diagnosis of PVS depends on a prior diagnosis of vegetative state (VS). A diagnosis of VS, in turn, is made on the basis of what can be observed externally. It is diagnosed by the complete absence over time of even the most minute signs of voluntary movement or responsiveness (Multi-Society Task Force, 1994, p. 1500). When, over the further course of time, it becomes clear that a patient in VS is not going to improve, the diagnosis of PVS is given.[3] Some patients initially diagnosed as vegetative do subsequently improve. However, it is important to emphasize that very few of those who are in VS for longer than 1 year improve significantly. Still, it not infrequently happens that patients who once showed no visible signs of awareness later begin to show signs of *minimal* awareness.

To account for this, in 2002, the Aspen Neurobehavioral Conference Work Group introduced a new diagnostic category, that of the minimally conscious state or MCS (Giacino et al., 2002). Like VS it is diagnosed on the basis of what can be externally observed. It differs from VS in that some minimal, typically inconsistent, signs of awareness are detectable at the bedside. But it is part of the diagnosis that communication with such patients via standard means is not possible.

Until recently, physicians felt confident making inferences about the interior life of such patients on the basis of these diagnoses. It was assumed that vegetative patients experienced nothing at all, and that minimally conscious patients experienced very little. But new research has raised interested questions.

Using functional magnetic resonance imaging (fMRI) technology, researchers have demonstrated that some number of patients diagnosed as either VS or MCS have regular patterns of brain activation in response to verbal commands—patterns that closely resemble the patterns of brain activation observed in healthy volunteers (Monti et al., 2010). The researchers first developed two mental tasks—a motor imagery task that involved imagining oneself swinging a tennis racket and a spatial imagery task that involved imagining oneself walking through the rooms of one's home and seeing all the familiar items—which they knew would engage different, highly specific areas of the brain. Having once established what the fMRI brain scans of healthy people looked like when they performed these tasks, researchers scanned the brains of VS and MCS patients while asking them to perform the same two tasks. Most patients were entirely unresponsive (showing little or no brain activity).[4] But the shocking result was that a few (all of whom had suffered traumatic brain injury) seemed to be following the researchers' verbal commands: Their scans revealed patterns of activation very much like those of the healthy controls. These were individuals who in some cases had been VS or MCS for years.

Going a step further, researchers decided to try to communicate with a patient using these fMRI techniques. Having established with healthy controls the possibility of communicating in this way, they placed one VS patient in the scanner and told him he was going to be asked a series of questions. He was instructed to engage in mental task 1 if the answer to the question was "yes" and to engage in mental task 2 if the answer was "no." In a still more amazing result, the patient seemed to answer the questions correctly.[5]

It is hard to know precisely what these results mean, but it is clearly both fascinating (from the standpoint of science) and disturbing (for it is possible that these patients have been suffering for years unbeknownst to us). Interpreting these results is made more difficult by the fact that the involuntary, unconscious part of the mind is far more sophisticated, and controls far more of our behavior, than most of us realize. This raises the difficult question of whether what these researchers have observed is a function of involuntary, unconscious processes or whether, more excitingly, it is a function of the voluntary, conscious mind. It is hard to see how we could hope to answer that question with any certainty on the basis of current evidence.[6] But even if the best we can say is, "They might be conscious," this is important. For it radically changes our understanding of these patients, undermining simple complacency about the nature of their experience. We can no longer be certain that they do not suffer, and this raises ethical questions about how we are to care for such patients.

My goals for this essay are twofold. First, I wish to articulate a general framework for thinking about the best interests of severely compromised incompetent patients such as patients in VS or MCS. In particular, I wish to articulate the two most important questions that ought to guide our thinking in this area, and to articulate as well some of the considerations that should go into answering these questions but all too often are overlooked or ignored. The two questions are simply these: "Do they suffer?" and "Are they deriving any benefit from their lives?"

My second goal is more substantive. I wish to defend an answer to the question of what would be best overall for patients in a PVS or a permanent MCS. I shall argue that, all things considered, it would be *better for them* to allow them to die. I limit my claims to these specific types of patients, because in these cases one does not have to struggle with the additional complicated question of how much weight to give to the possibility of future improvement. To grapple with that question would require some sense of how likely improvement is and also how great the improvement might be, factors that no doubt differ dramatically from case to case.

I also wish to acknowledge that my claim is based on assumptions about the feasibility of using certain types of technology in particular ways, and that if those assumptions are wrong or if they change, my conclusions might change as well. But the discussion to come will make it clear both what it would take for my conclusion to change and what it would change to. Moreover, although I am concerned about patients whose diagnosis is permanent, much of what I say is still relevant to thinking about the interests of patients who have not yet received that label. It is directly relevant to understanding their current interests, and it also helps us to frame what we need to ask about their future prospects. We would need some reason to hope that they might in the future improve to a point beyond minimal consciousness in order to conclude that their future interests provide us with reason now to continue their treatment.

I will assume throughout that we must view such patients as "incompetent" or as "lacking decision-making capacity." I treat these terms as interchangeable, purely technical terms (Kim, 2009, chapter 1). They refer to a particular level of decision-making capacity that our society has decided individuals must possess in order to be granted a certain normative status—that of self decision-maker. Competent adults in our society have this normative status. They are presumed to be in charge of their own lives, such that a competent adult's decision is the final authority when it comes to deciding what health care intervention he will receive.

Those who fall short of this standard must, of course, still be given excellent care and attention, but their currently expressed preferences (if they are able to express any) are not treated as authoritative. Final decision-making power lies in someone else's hands. It is worth emphasizing, because it is so frequently forgotten, that this does not mean that the person in charge of making decisions for an incompetent patient should neglect currently expressed preferences of the patient.[7] The person in charge of making a decision should consider and weigh everything that could possibly be relevant.

I nonetheless assume that the responses of VS and MCS patients in fMRI scans are not sufficient to license the conclusion that they are competent, leaving us to conclude by default that they are incompetent.[8] This matters, because some theorists have expressed the hope that we might be able to use fMRI technology and the yes/no communication task to ask these patients whether they are suffering or, even more dramatically, whether they wish to be kept alive in their current state (Monti et al., 2010, p. 589; Sinnott-Armstrong, 2011). But even if we eventually put such questions to these patients, and even if they respond, the mere ability to express a choice or preference is not generally considered sufficient to establish competency. In addition, a determination of competency usually requires showing that a person grasps the essential information, can relate that information to himself and his own situation (i.e., really grasp that it is relevant *to him*), and is not prone to obvious forms of logical error (Kim, 2009, Chapter 2). In short, we want to know a great deal about the quality of the processes—emotional and cognitive—that went into yielding the answer.

Moreover, it is extremely difficult to imagine how we could assess the quality of decision making for patients in VS or MCS. Yes/no answers are extremely limited in what they can reveal about a person's thought processes. This is not to deny that some day it may be possible to establish the capacity of some of these patients, but we are a long way from that now. Minimally, we would have to reach a point where we could communicate in a more open-ended way (i.e., moving past simple yes/no responses to questions) so that we could get a better sense of what the patient is thinking (if anything) independently of being prompted. For the time being, however, we must continue to view such patients as lacking decision-making capacity, and so we must continue to make decisions for them, in the same way as we do for other incompetent patients.

THINKING ABOUT BEST INTERESTS

The question that interests me is a question about what is good *for* these patients. First, however, I want to make clear the relationship between different bits of terminology used in different disciplines, remind readers of a few important distinctions, and delineate the guiding assumptions of my approach.

Bioethicists and lawyers tend to use the language of "best interests," whereas philosophers refer to "welfare," "well-being," or, in the most general case, "prudential value." The phrase "prudential value" is unfortunate in certain respects, for in ordinary language the word refers to adopting a self-protective attitude. Prudence in that sense requires adopting a general kind of cautiousness in one's dealings with others; in some cases, more specifically, it requires great care in the handling of money. Yet in its philosophical usage none of these associations are intended. Prudential value is just the kind of the value that self-interest seeks to realize, and the kind of value that good lives contain lots of. It is concerned most broadly with all that is or can be good for an individual.

Despite its limitations, I prefer the phrase "prudential value" because of its neutrality. The terms "well-being" and "welfare," in contrast, are explicitly concerned with the state of a person who is doing *well*, or for whom things are *good*. But human choices are concerned with more than just what is good absolutely (where this refers to being above some evaluative threshold). In cases where people are not doing well, we can still reasonably wonder which of the options open to them would be best overall. This is to wonder about which choice would be best from the standpoint of prudential value or, in other words, which choice would be good *relative* to the options available.

The relationship between good choices and prudential value can be complex for a number of reasons. To begin with, good choices must typically reflect an appropriate balance between concern for the present and concern for the future. A good prudential choice does not always maximize present value. Second, a choice can be good *intrinsically*, in that it directly increases prudential value in a person's life, or *instrumentally*, by bringing a person closer to being able to realize something of prudential value. Finally, it is important to remember that although we often speak loosely of "doing what is good or best for someone," this typically refers to making a good prudential *choice* for that person. In situations in which the option set is small and the possibilities grim, what counts as a good prudential choice may not be anything we would recognize as "good" without qualification.

How then is prudential value related to interests? As generally understood, interests are those things—events, persons, objects—that we have a stake in. A person is said to have an interest in something (e.g., an interest in obtaining or retaining a possession, an interest in having a certain relationship, an interest in the occurrence of a certain event) if that thing has the potential to affect the prudential value of his life. As this term is sometimes used in legal contexts, the potential to affect prudential value need only be slight, and in such cases the terminology of interests can come to seem fairly remote from the terminology of welfare.

However, as I use the term, the intended link is tight. For example, in order for it to be true that a person has a positive interest in the occurrence of a particular event, it must be true either that the occurrence of this event would directly and immediately improve his life *or* that it would greatly increase the likelihood of improvement by moving him significantly closer to being able to realize some good. So, for example, a person has an interest in eliminating or decreasing his suffering, because this improves his life in a direct way. Similarly, it is in the interest of a young person who aspires to go to college to prepare well for the Scholastic Aptitude Test. Preparation is no guarantee of success, but it greatly increases the odds, moving the student significantly closer to the realization of something prudentially valuable for him. Talk of *best* interests is simply talk of what would be best overall for a person once all of his various interests have been considered. The notion of best interests is thus equivalent to the notion of what would, relative to the circumstances, be the best prudential choice for a person.

For the purposes of this paper, there is no need to offer a theory of prudential value, and that is fortunate, for such theories are often highly controversial. Interestingly, although there is often great disagreement about what makes a life positively good, there is considerably less disagreement about what makes a life bad. In this sense, there is an important asymmetry in ordinary thinking about best interests, and in this case that asymmetry works to my advantage. I shall restrict myself, then, to two claims about how we should approach the question of prudential value for severely compromised patients such as those with disorders of consciousness.

My first claim is that if anything can tip the balance in favor of death (i.e., if anything can make it true that it is in a person's overall best interests to die), it is extended suffering. Hence, we must always inquire as far as we can about the suffering of such patients. The basic idea that suffering is a form of prudential bad and that its presence makes a life worse is not very controversial. I shall assume, however, somewhat more controversially, that a life of extended suffering—a life *dominated* by suffering—is not worth living. I say more in the next section about the nature of suffering.

Second, in order for continued life to be in the best interests of a patient, I assume that he or she must be deriving some sort of benefit from continued life. From a certain perspective, this may seem obvious. But the typical focus in medical ethics on pain and suffering can often lead us to overlook this fact. The mere absence of suffering, while necessary, is not sufficient to underwrite the claim that continued life is best overall.

These, then, are the questions that should guide our thinking. I now wish to elaborate more fully on each question in turn and consider what these questions can tell us, if anything, about the best interests of patients who are permanently in VS or MCS.

DO THEY SUFFER?

One of the more disturbing aspects of the recent fMRI-based findings is the worry that some patients previously thought to be entirely lacking in awareness might actually be suffering silently, internally, unable to reveal through any kind of voluntary movement what they are feeling and unable to receive any help. Discovering the existence and extent of suffering in these patients may be extremely difficult. Nonetheless, the question of suffering is clearly relevant in a way that means we must at least try to find out. Not only do we want to know whether they suffer, but, if they do, we want to know whether anything can be done to alleviate their suffering. And if their suffering cannot be alleviated, we need to remain open to the possibility that continued life may not be in their best interests.

But what exactly is suffering? What precisely are we trying to determine? Most people assume that the question before us is a question about pain. Clearly, pain is relevant, and if such patients experience pain, we should try to determine whether it can be medically treated.[9] However, I think the focus on pain is far too simple. Unlike many people, I draw a sharp distinction between pain on the one hand and suffering on the other.[10] In my view, "pain" picks out a certain class of physical sensations that occur in our bodies and which we find immediately and intrinsically aversive.[11] "Suffering," however, is more complex. Suffering is *affective*. It has to do with the way we interpret our immediate physical experiences, as well the events of our lives and the events occurring around us. It is often, although not always, a reaction to pain. But pain can exist without suffering (think of the pain experienced by athletes in the course of training and performing), and suffering can exist without pain (think of the overwhelming grief of a parent who has just lost a child).

When I say that suffering has to do with interpretations and with meaning, I am not claiming that suffering is produced by conscious, deliberative thought processes. It is not. Indeed, in some cases, the interpretation of the world that informs an individual's suffering can coexist with, and be at odds with, that individual's more consciously held views. This occurs because the interpretations that yield suffering are the product of the *affective* side our nature.

As psychologists use the term, "affect" is a general, all-purpose word for emotional phenomena. Emotions, as I understand them, are complex dispositions to feel, think, and act in certain ways. Sadness, for example, disposes us to certain physical sensations, leads us to think certain thoughts, and creates a tendency toward certain kinds of actions. Moods are also dispositions, ones that shape our intuitive perceptions of the world in negative or positive ways. In addition to emotions and moods, we can view individuals as having broader, more general dispositions that determine the frequency with which they experience particular types of emotions, as well as good or bad moods. I shall call these *affective* dispositions. A person with a set of negative affective dispositions tends to experience more negative emotions and is implicitly disposed to see the world through a dark-colored lens. Suffering, as I understand it, is the product of extremely negative affective dispositions.

Pain and suffering, although distinct, are of course intimately related. For we human creatures are constituted in such a way that extended pain invariably leads to suffering. Pain in the body forces attention to itself in a way that diminishes the rest of mental life. In some cases, we attach a fearful meaning to pain (e.g., as a sign that one is sick or dying). But even when an individual understands that his pain poses no threat to his health (e.g., that pain is felt because a nerve has been damaged but otherwise one's body is intact and free of illness), it may still be that the pain, if extreme enough and prolonged enough, will lead him to suffer, for the simple reason that constant pain restricts in numerous ways a person's ability to derive value and positive meaning from other aspects of his life. It is because of the suffering pain can cause that we sometimes view a life of extreme pain as not worth living. But although pain often leads to suffering, the more important point for the purposes of this essay is that suffering can exist *even in the absence of pain*. An individual will almost certainly suffer, for example, if most of what he once cared for has been lost and he sees no hope for anything positive in his future.

When considering the question of best interests, courts have generally been open to the idea that a life of extended suffering may not be in the interests of an incompetent patient. However, they have typically understood suffering entirely in terms of pain. For example, in the famous case of Claire Conroy (*Matter of Conroy*, 1985), the Supreme Court of New Jersey articulated two best interest standards, both focused on pain, that it said could be used for making life-and-death decisions for incompetent patients in the limited case in which there was insufficient evidence of what the patient himself would have wanted. The first test, which the justices labeled the "limited objective test," required decision makers to be certain that the burdens of continued life with treatment (where burdens are understood in terms of pain that cannot be eliminated with treatment) outweigh any benefits the patient is getting from his life, and also that there be some evidence that the patient would have wished to forego the treatment in such circumstances. According to the second test, the "pure objective test," life-sustaining treatment may be stopped in the complete absence of evidence of a patient's wishes if "the recurring, unavoidable, and severe pain of the patient's life with the treatment would be such that the effect of administering life-sustaining treatment would be inhumane."

In short, in the view of the majority in that case, it is only when a person is in pain, and the pain is as severe as that described, that we can be confident that it is in a person's best interests to die. This strikes me as problematic in at least two ways. First, if one can be certain that the patient is not benefiting from his life (the topic of the next section), then there is no reason to insist that pain be so extreme. In such a case, the presence of chronic pain at any level should suffice to tip the balance. More importantly, however, there are forms of suffering that this account completely overlooks.

It may be that in the *Conroy* case, which involved a woman in the final stages of dementia, the justices whose views are reflected in the majority opinion assumed that the patient was not mentally intact enough to suffer, even if she could experience physical pain. Moreover, it is likely true that the possibility of suffering as I have described it increases as the level of internal mental capacity increases. Still, it seems that one of the lessons to be drawn from the recent fMRI studies is that we cannot always accurately predict interior life on the basis of exterior signs. If that is right, then we should be cautious about assuming that such patients do not suffer. We should instead admit that in most cases, we simply do not know.

When we consider the specific case of the five VS/MCS patients who were able to carry out simple mental tasks during fMRI, it is hard to know what to conclude. We do not know whether they are aware enough to think much, and we have no way of knowing what they can think about. Just because parts of the mind have been preserved, we still do not know whether these patients can remember much of their previous life, whether they experience emotions, whether they contemplate their current plight. But it does seem safe to assume that *if* these individuals are capable of thinking about and evaluating their current situation, they are most likely suffering in the extreme. For they have lost a great deal, have little or nothing to gain, and are isolated beyond imagining. Nor is it feasible, given the current costs, to suppose that fMRI will offer such patients a way to reconnect with their world. At best, right now, it may offer a few brief opportunities for limited exchange. Such a life is hard to imagine. However, because we do not know whether this is their life or not, let me turn for the moment to discussion of the second guiding question.

DO THEY BENEFIT FROM LIFE?

Even in the absence of suffering, if we are to say that it is in a person's *best interests* to continue living, we must establish that life is *beneficial for him*. He must be deriving some benefit from his life.[12]

It can be easy to overlook the significance of this question, in part because theorists often frame the issue in terms of what it would take for it to be true that *death* is in someone's interest. But even if we grant that extended suffering makes death preferable to life, it does not simply follow that when such suffering is absent, life is thereby preferable to death. This way of thinking overlooks the fact that there are conditions an individual can be in where it is neither good for him to live nor good for him to die, where interests do not point one way or the other. The traditional understanding of PVS as the permanent loss of all capacity for awareness is a case in point, as is anencephaly in infants. When there is no capacity for awareness, continued life offers no opportunities for either benefit or harm.[13] One of the disturbing aspects of the recent fMRI findings is that we can no longer be sure which patients fall into this category. But we need to at least consider what would make continued life beneficial: What are the minimal capacities needed for being able to derive benefit from one's life?

In thinking about this issue I have found an older debate—a debate about when we ought to treat severely impaired newborns—illuminating. I do not mean that the issues are exactly parallel or that the two patient populations can be usefully compared, for they cannot. But I hope to put to use in a novel way some of the insights from this earlier discussion.

To understand this older debate, we need to remind ourselves of a dramatic shift in thinking that occurred in the 1960s and 1970s about how to make decisions for severely impaired newborns.[14] For centuries, majority opinion deemed handicapped lives to be not worth living. Although there are, no doubt, many factors that contributed to this opinion, it largely reflected the judgments of ordinary people that it would be terrible to become handicapped. In other words, people with species-typical physical and mental capacities tried to imagine what their life would be like were they to lose those capacities, and they concluded that it would not be worth much.[15] They did not, by and large, consider the nuanced question of whether limited capacities might still be better than none. Nor did they typically invest much effort in trying to imagine the perspective of someone who from birth had never possessed the full set of species-typical capacities. Because these prejudices were so dominant, the question of whether to attempt to save the life of a child born with disabilities usually turned on whether the parents were able and willing to care for the impaired child. Rather than actually wrestling with the difficult question of whether a disabled child might, despite the handicap, have a reasonable chance at a decent life, people tended to let parental burdens settle the matter.

This attitude radically changed in the 1970s and early 1980s. When it did, new criteria for making treatment decisions for impaired newborns were needed. It was clear that some severely impaired children simply could not survive or could have only a short, miserable life. Hence, even though many agreed that the old system was bad, it was also clear to many of those involved that we should not simply try to save every child no matter what their condition.[16] Instead, it was suggested that we should be in the business of saving children whenever (as best we could tell) it was in a child's best interests to be saved.

As in other discussions of best interests, there was a general consensus that a life composed entirely of suffering could not be worthwhile.[17] But beyond that, consensus seemed to end. Some theorists maintained, and I would agree, that a child who is incapable of consciousness has no interest in continued life. For the anencephalic infant, as for an adult who has truly, permanently lost the capacity for consciousness, life is neither beneficial nor harmful. But what about those children who are born with severe cognitive defects but who have some small degree of consciousness?

The standard move at the time was to assume that as long as a child is sentient but not suffering, it is in the child's best interest to live. Some degree of awareness, however small, seemed to mark the boundary on the other side (i.e., the point at which life becomes a benefit). This view was defended, for example, by the legal theorist John Robertson in an influential essay from 1975 (Robertson, 1975, p. 269). Robertson thought that it was clearly in the best interests of many—and probably most—mentally handicapped children to live. But he recognized that there can be a real question about the interests of certain extremely impaired but nonetheless sentient children. He gave as his example the case of a "profoundly retarded, nonambulatory, blind, deaf infant who will spend his few years in the back-ward cribs of a state institution" (p. 254). We are to imagine that such a child's mental handicaps are severe enough that his conscious experience is at best highly fragmented. It is completely unclear what, if anything, he understands about his world.

Although recognizing such cases as the tough ones, Robertson ultimately concluded that it is in the best interest of such a child to live. However, his argument seems to boil down to the claim that it is in the child's best interest to live *because there is no harm for the child in such a life*. However, this confuses having a reason to live with lacking a reason to die. As mentioned earlier, it is possible to lack reasons of both sorts. It seems hard to imagine that a child as profoundly retarded and deprived of sensory input as the one he described could be deriving benefit from his life. Mere sentience is not in itself sufficient for deriving benefit from life. Although Robertson is right that life is not a harm to the child, he has not shown that it is *good for him* to exist.

In an insightful essay from roughly the same time period, John Arras (1984) took issue with Robertson's conclusion. He clearly thought it a mistake to conclude that we should save children who are this severely impaired. But, unlike me, he was willing to simply cede the language of best interests to those who, like Robertson, claim it is in the best interests of such a child to live. In effect, he granted that it is in the child's interests to live, but argued that if that is so, we must move "beyond best interests" (p. 31). Although I agree with Arras' conclusion about the case, I think we should retain the language of best interests and say, instead, that it can be in the best interests of a child to live only if the child either has, or will in the course of normal development acquire, the capacity to derive benefit from living. This would still support the practical conclusion Arras favors—the conclusion that we should not treat such severely impaired infants. For I assume that when treatment is neither beneficial nor harmful to an individual, then it is perfectly legitimate to allow other factors to help determine the decision, even factors that would not normally be allowed to have weight.[18] In short, if a child cannot derive benefit from life, then it makes sense to consider the financial and emotional burdens of looking after the child and perhaps to decide that it is simply not worth it. Such a decision will not be detrimental to the child, because we have already stipulated that we are dealing with cases in which death is not a harm to the child.

Arras went on to consider under what conditions we should try to save impaired newborns. If mere sentience is not the important factor, what is? My suggestion is that we view this as a question about the minimal capacities necessary for life to be good for the individual whose life it is (or alternatively, the minimal capacities necessary for an individual to be able to extract value from his own life). In what follows, I want to consider two possible answers, the second of which is embraced by Arras.

The first answer is suggested by the way the question was framed. If life is valuable for a person only when he is capable of deriving value from it, then minimally it seems that the individual must be a *valuer*: a being capable of valuing. The notion of valuing has been understood in a number of different ways by different theorists. Philosophers, in particular, have tended to think of the mental capacity to value in highly intellectual terms, making of it a more complex capacity than what we need here. For example, some have suggested that in order to value something, one must view oneself as correct in valuing that thing.[19]

The notion of *caring* may be more helpful. Caring, as I shall understand it here, is a simpler capacity, one that we possess before we become valuers and one that we may sometimes retain even though we cease to be valuers. For example, I assume that small children and many higher animals are capable of caring about things, even though they lack linguistic capacity and the more sophisticated cognitive commitments philosophers associate with valuing.[20] Minimally, to care about something, I must be capable of distinguishing it from other things in the world and (assuming the object of care is literally an object—something or someone in my environment)[21] capable of perceptually tracking it to some degree. I must also, of course, recognize the object of my concern over time. And this is not all. To care about X is to have certain, very primitive emotional dispositions vis-à-vis X. For example, I will be pleased or happy when I perceive X; I will want X to remain with me; I will be sad when X disappears; and so on. These simple capacities can be in place despite an individual's inability to articulate any of these feelings, and despite his having very little understanding of the person or thing he cares about.

I shall refer to this first answer as the answer *in terms of caring*. The idea is that when no suffering is present, a being with the capacity to care has at least some positive interest in living. An infant may lack this capacity at birth, but if he will develop it in the normal course of development, then it is in his best interest to live. However, beings permanently lacking this capacity are, like the anencephalic infant, neither benefited nor harmed by continued life, despite having some minimal degree of awareness.

The second answer was originally proposed by the Catholic theologian, Richard McCormick (1990),[22] and later taken up and endorsed by John Arras in the same essay discussed earlier (1984, pp. 32–33).[23] Call this the answer *in terms of relationship*. In this view, continued life is good only for those impaired infants who have (or will have in the course of normal development) the capacity for relationships with others.

A lot turns on what exactly we take a relationship to be. Just as it is possible to construe valuing in a way that makes it too complex for our purposes, it is similarly possible to describe the necessary requirements of genuine relationships in a way that is too complex. For our purposes, we should seek to describe the simplest possible form of relationship, and it is useful in this regard to consider the change that occurs in infants at about 3 months of age when, as people say, "social smiling" begins. Before that point, infants smile in random ways, almost as if they are just trying out their facial muscles. Of course, for all we know they might be smiling in response to pleasant feelings. There is no way to be sure. But even if that is so, it is not clear what significance this would have. For it is doubtful that such responses would be evidence of either the capacity to care or the capacity to engage in relationships.

Once an infant begins to smile socially, however, he begins to develop very simple relationships with other human beings—usually his mother and father. He grasps in the simplest sense that other people are distinct from him; he recognizes and tracks particular people perceptually; and he responds to them in distinctive ways. The most minimal form of communication can begin to occur at this point, because the infant can share very simple feelings with the other person and learn in a very crude way about the feelings of others. This is not to say that he has any comprehension of what the feelings of others really are or mean. It is just to note the very simple fact that he smiles for their benefit (to see how they will respond) and looks for smiles in return. He grasps in a very simple way that certain gestures and looks are intended for him. Although by no means the richest of human relationships, this is, I think, a relationship. And although a life with no more in it than this would indeed be impoverished by our standards, it nonetheless seems plausible that in the absence of misery or suffering from other causes, a person with at least this level of capacity could extract something good from his own existence in a way that someone lacking such capacities could not.

Both of these answers—the answer in terms of caring and the answer in terms of relationship—contain important insights. In the case of impaired newborns, the capacity to care may come closer to capturing what we really think must be in place before life is beneficial, but the capacity for relationship may be easier to translate into some sort of concrete standard. Moreover, it is plausible that in developing children, the two capacities typically arrive at a similar point in time and serve to reinforce one another during the course of future development. The answer in terms of caring may also shed some light on why we think the capacity for relationship is so important, because it is only through sustained interaction with our world that we find an outlet for our caring. The capacity to interact with our world is necessary both so that we can locate objects to care about and so that, once we care, we can follow the objects of our concern. In short, without the capacity to relate to the world, the capacity to care would, at best, offer us nothing, and might in the worst case be simply a deep source of frustration.

Of course, our primary concern here is with the best interests of patients who have disorders of consciousness. What should we say about *them*? First, I think we should conclude, much as Arras did in the case of impaired newborns, that more than mere sentience is necessary for life to be a benefit. Second, I think that the two answers canvassed in the discussion of newborns—the answer in terms of caring and the answer in terms of relationship—can help resolve the question of whether patients with disorders of consciousness are able to benefit from life. But the way in which they may help is not immediately obvious.

The problem we face with these patients is that we do not know how mentally intact they are, or what their interior mental life consists of. We are, instead, in the extremely frustrating position of knowing only that previous estimates of their mental life were, in a small number of cases, deeply mistaken. We now know that some such patients have more mental life than we previously thought, but that is not really saying very much. Nor are we in any position to know whether these patients have the capacities I mentioned in the discussion of newborns. Because they were previously competent adults, we know that they *once* possessed the capacity to care and the capacity for relationships. But we have no way of knowing to what extent those capacities remain intact and functional.

Nonetheless, despite our ignorance, there is one significant piece of information we possess that is relevant here. Whether or not these patients have the capacity to care, current resource limitations ensure that they cannot in their present state interact with their world or sustain meaningful relationships with others. The research suggests that they have the capacity to hear, but mere passive reception of sensory input, without the ability to respond to it, is neither sufficient for relationships nor an obvious benefit in its own right; it may instead simply be a source of deep frustration. Indeed, reflection on the case of patients with disorders of consciousness suggests that the capacity for relationship should really be thought of as a *set* of capacities. Some of these can be considered as "internal" mental capacities, such as the capacity to focus one's thoughts on another person or object. For all we know, these kinds of capacities *may* be intact in these patients. But the capacity for relationship also requires the ability to communicate, if only in the most primitive ways. And, outside of the brain scanner, these patients show only intermittent signs of awareness and have no ability to communicate.

Nor do the fascinating results of fMRI technology suggest that this situation will change, at least not soon. It is not now, and is unlikely to be at any time in the near future, feasible for these patients to sustain relationships with others via fMRI. I shall return later to the issue of what it would mean if that were to change—if it were to become possible to sustain relationships with such patients via technology. But for now, because relationships are not possible, it seems clear that such patients are incredibly isolated. Given just how isolated they are, if they are mentally intact, it seems likely that they suffer and suffer greatly.

Therefore, it seems that despite our ignorance of the interior life of these patients, we can conclude that one of two things is true of their current state. On the one hand, it *may* be that the mental life these patients experience is so broken, fragmented, and confused that they lack the capacity to focus on, or care about, much at all. If that is the case, my view suggests that currently they are not deriving any benefit from life. If that is their permanent lot, then they never will derive benefit. In such cases, continued life is neither in the patient's interest nor against it. On the other hand, it is also *possible* that such patients may be more intact mentally than even the results so far have suggested. But even if that is the case, they are still extremely isolated. In these patients, too, because they cannot interact with their world or sustain even the simplest relationships with others, it is also likely that they do not benefit from life. But here, unlike the case of impaired newborns, we should not conclude that continued life is neutral, neither good nor bad. For *if* in this case the lack of capacity for relationship is partial—if it involves the lack of any ability to communicate coupled with fairly sophisticated interior awareness—then it is likely that such patients suffer.

Some will no doubt take issue with my inference from isolation to suffering and argue that we cannot *know* that these patients suffer (or that they would suffer if intact enough to be aware of much). Although I will certainly grant that we can't know for certain, I do think it is the most plausible view, and the kinds of considerations sometimes thought to cast doubt on this notion do not really do so. First, it is sometimes said that ordinary people vastly underestimate the quality of life of those who are severely disabled and that, to the surprise of those who are "normal," many such individuals wish to remain alive.[24] That such mistakes are easy and frequent, I fully grant. Indeed, I usually find myself on the side of those who argue for a more fine-grained appreciation of the experience of those who are disabled, and I favor maintaining the life of most disabled people, even many of those who are severely cognitively disabled.[25] However, most disabilities, even extreme ones, do not leave individuals so isolated. Communication of some sort is typically possible, and such individuals can thus derive benefit from the love and care of others.

Second, it is sometimes pointed out that even patients with locked-in syndrome (LIS), a neurological disorder in which a patient is fully conscious despite being almost entirely unable to move, often desire to live (Doble, Haig, Anderson, & Katz, 2003; Laureys, Pellas, & Van Eeckhout, 2005).[26] Surely, this suggests that we should not be too quick to judge in the case of the minimally conscious. However, as I see it, this claim underscores just how poor we really are at seeing the differences and similarities that *matter* from the standpoint of quality of life. From our point of view, it may make sense to compare a patient with LIS with one who is minimally conscious, because both are completely bed-bound and completely dependent on high-level care, and because in both cases it is hard for us to know what is going on "inside." But there is a difference here that makes all the difference. Most individuals with LIS are able to communicate, even if communication is laborious and burdensome.[27] Locked-In patients can thus maintain relationships with those around them. But when there is no way to interact with one's world and no way to communicate with people, I think that emotional suffering will almost certainly be present. And it is not likely that we can eliminate such suffering, because we cannot (at least now) really address its root causes.

To summarize, if we focus simply on what we know about the current interests of these patients, we can reasonably assume that either life is neutral (neither beneficial nor harmful because the patient's consciousness is too fragmented), or it is negative (because the patient suffers). Thus, despite our inability presently to tell which is the case, we should conclude that it would be better overall to allow such patients to die. Assuming that things will remain as they are, death is either a neutral event or a great blessing.

However, this brings us back to the question of whether this state of affairs really is permanent. Even assuming that we restrict our attention to those patients for whom the *diagnosis* is permanent (i.e., they are permanently in VS or permanently in MCS as currently defined), there remains a question about whether relationships might one day be established with such patients via fMRI or some simpler technology. Indeed, one very interesting study has shown that electroencephalography (EEG) can be used to detect limited awareness in VS patients, in much the same way that fMRI has been used (Cruse et al., 2011). To date, this method has been employed only to test for patients' ability to follow simple commands (e.g., "Imagine you are squeezing your right hand into a fist and then relaxing it") and not for communication purposes. But perhaps it will eventually provide a way to communicate. Because EEG is so much less expensive, it is at least more feasible to think that this method might become widespread. *If* that came about, then I would, other things being equal, favor seeing whether it is possible to sustain meaningful relationships with such patients, relationships that would make their lives worth living and eliminate their suffering. However, much remains to be seen.

Two further cautions are in order. First, we should be careful not to fall into the trap of assuming that if communication is possible, sustaining life must be overall good for the patient. In my view, the possibility of relating to the world is a necessary condition of being able to derive benefit from life, but it is not a sufficient condition. Minimally, suffering must be absent as well. Although it seems plausible to assume that a completely isolated life would be a life of suffering (if the patient were mentally intact enough), there is no guarantee that suffering would disappear once some kind of interaction is established. This would depend on many factors. We can, for example, imagine a case in which even though we are able to communicate with a patient in a primitive way, the gap between what the patient is capable of thinking and what he can actually express might be intolerably large from his point of view. So, he might *still* suffer. Also, it may matter greatly how frequently the patient is able to communicate with others and whether he is able to communicate regularly with the particular people he wants to communicate with. It will presumably also matter how temperamentally able the patient is to find new focuses in life and adapt to his position. People differ greatly in this ability. *If* we ever reach this stage, the answers to such questions will no doubt vary from individual to individual. So, even though life without relationships is not worth living, we must not assume that once relationships are possible, continued life is a pure benefit.

Second, and perhaps even more importantly, I am concerned about the conclusions some people may draw about best interests, given that right now such communication is not possible for the majority of patients in PVS or permanent MCS. I am worried that loved ones or courts may, with the best of intentions, seek to sustain the lives of such patients indefinitely in the hope of one day being able to re-establish communication. There is a common tendency to assume that future benefits or goods can justify almost any amount of current suffering or pain. But that is simply not true to lived human experience. A certain limited amount of suffering may be worth enduring if it is the only way to get to a future point at which great goods await us. But the goods in the future must truly outweigh the current negatives, and in many cases of extreme suffering, it is highly unlikely that the future can really redeem the present. I am thus not convinced that it would make sense for a family to decide today to continue treatment in a loved one because they hope that in the next few years fMRI or EEG may be available. Unless that family is in the position of knowing they will be able to try such a project within the very near future, within at most a few months, I think they risk leaving their loved one in a state that may be horrible—and doing so for no clear future benefit, or for a benefit that, while real, cannot justify the suffering that came before. We should thus exercise great caution in our judgments about these cases.

This last point is extremely important, because some theorists seem to have concluded (in light of the recent research findings) that things are looking up for PVS and MCS patients. But really, the picture has simply become much more complicated in a way that is deeply worrisome. Either such patients are beyond prudential value (in which case not much has changed, because that is what we used to assume), or they are capable of being benefited or harmed and may actually be suffering quite a bit. Although I will not deny that there are cases where it may make sense to try to establish some sort of connection with patients via technology, we should proceed very cautiously and hesitate to do anything that might prolong suffering. We should always keep in mind that despite our epistemic limitations, we know that currently things are either neutral for them or bad. So, the default assumption—which we could override, but which we should not override without carefully worked-out justifications—should be that it is best for these patients to allow them to die.

CONCLUSION

I have argued that when we think about best interests, we need to consider not only the question of whether a patient suffers but also the question of whether he is getting anything positive out of his life. I assume that a life of extended suffering is not worth living, and that when the answer to the first question is a clear yes, we ought to allow the incompetent patient to die. Moreover, I think it is important to remember that suffering can be present even when pain is not (although we shouldn't forget about pain). Emotional suffering can do just as much to undermine the value of life as suffering caused by physical pain. Unfortunately, in the cases that interest us here, it is extremely difficult to know whether the patients suffer.

I have also argued that we should ask whether a person is deriving any benefit from his life. If he is, this must be considered and weighed against any burdens. There will be a category of patients for whom life is neither beneficial nor harmful. This is true, for example, of patients who have permanently lost all consciousness (although, as the recent research reveals, with certain classes of patients it is no longer so easy to determine when that has occurred). Nonetheless, when there is no conscious awareness, life is neither beneficial nor harmful. But the same also holds for some conscious patients, namely those who are cognitively incapable of caring about anything or of forming and maintaining even the simplest sorts of relationships.

Finally, I have argued that in the case of patients in PVS and permanent MCS, we may be able to get around some of the difficulties that arise from the fact that we have so little knowledge of what goes on inside. For even though it is not clear what precise level of awareness such patients have, it is likely that one of two things is true currently: Either they are in a mental position such that life is neither a benefit nor a burden, or they are more cognitively intact and thus able to be benefited, but because of their extreme isolation and inability to interact with anything or anyone they care about, their life is a life of suffering. They may have the internal capacities for forming and maintaining relationships, but they have no way to operationalize these capacities. In that case, life becomes a burden. Thus, if we limit ourselves to reflection on the current situation of such patients, best interests point toward allowing them to die. Death is either neutral or a blessing.

This conclusion must be qualified, of course, because it is based on claims about the importance of relationships and because it is at least possible that some of these patients may, at some point in the future, be able to maintain relationships. However, we must be extremely cautious here. Many issues unrelated to my topic of best interests (e.g., social policy questions about cost and access) will have to be solved before it even becomes possible to try establishing relationships with such patients via technological means. Moreover, we should be very cautious about assuming that the improvement of being able to communicate will make life seem worth living for such patients. That will depend on many, highly individual variables. Nor should we rush to conclude that we ought to maintain such patients indefinitely until such time as it becomes possible to try communicating with them. Not all suffering can be compensated by future gains, and we must take seriously their current situation.

NOTES

1. The abbreviation "PVS" can be used for either *persistent* vegetative state (a vegetative state that lasts longer than 1 month) or *permanent* vegetative state (an irreversible vegetative state, which is usually diagnosed after 1 year in a persistent vegetative state). For a description of the difference, see Multi-Society Task Force (1994, p. 1501). Because it is the *permanent* vegetative state that interests me most, I will use PVS to stand for that.
2. Some patients also end up in vegetative states in the final phases of degenerative or metabolic neurologic diseases, and some are vegetative as a result of developmental malformations of the nervous system (e.g., anencephalic infants). For a detailed list of the causes of vegetative states, see Multi-Society Task Force (1994, p. 1503).
3. A persistent vegetative state is labeled "permanent" when "the chance that the patient will regain consciousness is exceedingly small" (Multi-Society Task Force, 1994, 1501). The Task Force suggested that, for brain-injured patients, this point is at 1 year. Others, however, have disagreed, drawing a different conclusion from the same sample of 434 patients studied by the Task Force. Of 65 patients still alive and vegetative at 1 year, 7 patients (10.6%) went on to regain some degree of consciousness *after 1 year.* These authors contend that it is hardly fair to say the state is "permanent" at 1 year when 10.6% still go on to improve (see Borthwick, 1996; Stone, 2007). Despite this dispute, I assume there is *some point* at which we can conclude with high certainty that further recovery is extremely unlikely, even if the 1-year mark is not it. And I will assume for the purposes of this essay that "permanent" refers to patients who have passed that point, whatever it may turn out to be.
4. The study included fifty-four patients, thirty-one with a diagnosis of MCS and twenty-three with a diagnosis of VS. Patients varied considerably in the original cause of their disordered consciousness. Of the total, five patients responded to the motor imagery task, and four of them also responded to the spatial imagery task. Thus, out of the original fifty-four patients, roughly 9% responded. However, it is significant that all of the responders had suffered traumatic brain injury. There were thirty-three traumatic brain injury patients in the study, making the response rate *within that group* roughly 15%. This is important to keep in mind because, as of now, there is no reason to rethink our original understanding of the state of those patients who are vegetative or minimally conscious as a result of loss of oxygen, degenerative illness, malformation, or other nontraumatic causes. The findings appear to be applicable only to traumatic brain injury.

5. The patient who was given a "communication" scan had originally been diagnosed as VS, but after the study, he was rediagnosed on the basis of bedside observations as MCS (Monti et al., 2010, p. 583).
6. As other selections in this volume attest, I am not alone in being unclear about precisely what these results mean. Over the past 30 years, research psychologists have documented the astounding number of ways in which the fast-working part of the mind—what Daniel Kahneman (2011) has called our "System 1"—can process information and develop sophisticated responses without our even being aware of its operation. "System 2" is the label reserved for the conscious, voluntary part of the mind, which often adopts the suggestions of System 1. The variety and complexity of examples leaves me wondering whether with fMRI we are listening to a patient's System 1 or System 2.
7. For just one of many possible examples, see Steffen and Franklin (1985, p. 13). They write: "Is Mr. B. competent to give or withhold consent to treatment? This is the central question, for if it is answered 'no' then the answer to the first question—how vigorously the staff should question him—is 'not at all.' It would be pointless to push Mr. B. for answers only to discard them as the views of an incompetent person." The assumption is that the views of an incompetent person are irrelevant to treatment decisions, which in my view is too strong.
8. In ordinary cases, adults are presumed to be competent until they show signs of incompetency, at which point an assessment must be conducted. However, because these patients have suffered severe brain injuries and have been unable to communicate until now, the presumption is reversed, and we can and must assume that they are incompetent until we have sufficient evidence of competency. On the presumption of competency, see Buchanan and Brock (1990, Chapter 1).
9. On the topic of pain, see also Valerie Gray Hardcastle's entry in this volume, "Minimally Conscious States and Pain: A Different Approach to Patient Ethics." She and I agree that it is highly relevant whether or not a patient is in pain, and we agree that much more should be done to try to treat pain in such patients. Hardcastle has her own distinctive reason for thinking that we ought to treat all VS and MCS patients for pain, although I am not sure I am convinced by it. However, the question is overdetermined, because there are already sufficient reasons for thinking that we ought to be doing more to treat pain in VS and MCS patients.
10. Eric Cassell, whose work has greatly influenced my own, makes a similar distinction between pain and suffering. Like me, Cassell (1982, 1991) sees suffering as being about *meaning*. However, my own, more detailed account of suffering differs from his.
11. Observant readers will note that this definition of pain requires pain to be conscious, signaling a difference between Hardcastle's use of the term "pain" (Chapter 12 in this volume) and my use of it. She uses the word "pain" to refer to specific brain processes that underlie certain aversive experiences, whereas I use it to refer to those experiences directly. Her view allows her to say that pain is present whenever these processes are occurring, even if the subject is not conscious. I would simply say that pain processes can occur whether or not the person feels pain. Either way, an interesting question arises, which we would each state

slightly differently. She would ask, "Does pain matter when it is not felt?" I would ask, "Do pain processes matter when they do not produce felt pain?" Hardcastle says "yes" for reasons having to do with the brain damage caused over time by untreated chronic pain. Although I would agree with her for most patient populations, I remain unconvinced for the special case of patients in PVS or MCS.

12. A nice example of an instance in which theorists do insist that we ask this question is found in Dresser and Robertson (1989, p. 240).
13. (Warning: This is a footnote for philosophers. Everyone else can feel free to ignore it.) Some philosophical readers may object that this assumes the truth of some sort of mental-state theory of prudential value as opposed to a state-of-the-world theory (Griffin, 1986, p. 17). Among philosophers, at any rate, such theories are highly controversial, so if it were true, that would indeed be problematic. However, no such assumption is being made. The standard example of a state-of-the-world theory is the desire theory, according to which what is good for someone is getting what he wants, here understood as *the coming true of his desires*. Because desires can come true without our knowing it, the theory implies that we can be benefited (or harmed) without our knowledge. What people sometimes fail to see, however, is that even with a state-of-the-world theory like the desire theory, it is unlikely that life would benefit a permanently unconscious patient or that such a patient could be said to have an interest in continued living. A permanently unconscious patient forms no new desires, and his old desires fall into one of two categories: Either they have already been frustrated or satisfied (e.g., the desire of a patient to some day swim the English Channel, was frustrated on the day he entered PVS), or they remain open, in the sense that future events in the world may either satisfy or frustrate those desires—but this can happen equally well whether the patient lives or dies. (For example, suppose a patient, while competent, had worked hard to raise awareness of the importance of literacy and founded a program to improve literacy in the schools of her city. She hoped that one day the schools would embrace her program and incorporate its fundamental principles into the basic curriculum. At the time of her injury, this desire had not yet been satisfied. If it were to happen while she is in PVS, desire theorists would say that this is good for her; it adds value to her life, even though she is unaware of it. But they would say *the exact same thing* if it happened after she died.) In short, state-of-the-world theorists think that a permanently unconscious person retains a number of interests, and they object to the common assumption that beings lacking the capacity for consciousness have no interests. But my claim is not that such beings lack interests; rather, but they lack *an interest in continued life*. And in the vast majority of cases this will be true precisely because the interests that remain do not typically depend in any way on whether the patient lives or dies.
14. For a detailed historical overview, including brief descriptions of many significant cases, see Pence (2008, Chapter 8).
15. Psychologists know that individuals are not, generally, good judges about such matters. For example, healthy individuals typically assume that they would be much less happy living with a particular disability than, in fact, most people are who actually do live with that disability. Although there remain complex debates about how best to understand the existing data, it seems safe to

say this much: Individuals with even severe disabilities frequently value their lives greatly, and typically much more than nondisabled persons suspect. (See Albrecht & Devlieger, 1999; Brickman & Janoff-Bulman, 1978; Damschroder, Zikmund-Fisher, & Ubel, 2008.)

16. There were, of course, exceptions. The Reagan Administration's policies famously rejected any quality-of-life judgments and insisted on saving every child. Moreover, in a few cases, the administration's "Baby Doe squads" demanded extraordinarily invasive treatment that, in the end, gave an infant only a few extra days of life (see Pence, 2008, p. 202).
17. My discussion of newborns may strike some readers as overly simplistic because I am focused on only one aspect of decision making. It is important when thinking about newborns to distinguish between questions about what capacities a child must have if he is to derive benefit from his life and, on the other hand, *epistemic* questions about when we can be certain that those capacities exist or will be likely to exist. Given how extremely difficult it is for physicians to predict what a newborn child will be capable of later on, epistemic considerations will often lead us to err on the side of treatment, particularly if the child is not suffering. I think that both questions (about capacities and about our ability to detect them) are important, but in the context of this chapter, in which it is the question of capacity that has relevance for my larger topic, I focus on that.
18. A similar point was made by Dresser and Robertson (1989, p. 240).
19. For example, Agnieszka Jaworska (1999, p. 114) stated that part of what it is to value something is to think that it would be a loss to oneself if one were to stop valuing it. And in a later paper on caring, she explicitly argued that valuing (but not caring) requires seeing oneself as *correct* in one's attitudes (Jaworska, 2007, p. 541).
20. Jaworska, like me, uses the word "caring" to denote something simpler than valuing, an attitude that can be had by individuals who lack some of the more sophisticated cognitive requirements necessary for valuing. However, two points are in order. First, although Jaworska uses "caring" in a simpler sense, not all philosophers follow this example. Some construe caring in ways that make it almost as complex as valuing (see, for example, Frankfurt, 1989). Second, even Jaworska uses it in a way that makes it more complex than I wish it to be understood here. I allow that young infants and many higher animals *care*, whereas Jaworska (2007, p. 564) is explicit that, on her account, infants and animals do not care, but 2-year olds and patients in the middle stages of dementia do care. Ultimately, I may need to find a third term to put to use for my own purposes, but for the moment, it is sufficient that the reader understand caring in terms of my actual description of it here.
21. Caring itself does not literally require these kinds of perceptual and tracking capacities, but they would be necessary for one to begin to care about an object or a person in one's environment. And infants would need such capacities to get started as carers. However, more cognitively sophisticated beings might care about ideas or other abstract things, and for that, perceptual capacities and tracking capacities would not be necessary.

22. McCormick saw the relationship standard as deriving from Christian theology. He argued that life is a relative good given to us by God for the pursuit of certain spiritual ends, namely, the love of God and neighbor. He argued that it is through our love of *others* that we encounter and come to love God. Hence, "the meaning, substance, and consummation of life are found in human relationships and the qualities of justice, respect, concern, compassion, and support that surround them" (McCormick, 1990, p. 30). Because he envisions relationships as a way to grow morally and engage with the divine, it has always seemed to me that he must have a rather sophisticated notion of relationship, one that would demand more in the way of capacities than my own account would. So although I find his initial idea appealing, I am willing to adopt it only in a severely altered form. Moreover, I do not think it needs religious underpinnings, for it is independently plausible to think that the capacity to engage in relationships is a minimal requirement for being able to derive benefit from life.
23. How Arras intends to understand the nature of relationships, however, is not fully clear.
24. For example, this argument was used by Jim Stone (2007, p. 89) to support the idea that we should maintain patients who are minimally conscious. However, the kinds of "disabled" individuals discussed by Stone to make his case are in a very different category from those who are minimally conscious. Although they are cognitively disabled (he described his own sister, a stroke victim, who could not "read, write her name, or count past three"), they are able to communicate, relate to others, and derive enjoyment from their lives. I am thus not persuaded that their case is at all relevant to those involving disorders of consciousness.
25. Again, my view implies that we need a way of distinguishing within the class of severely cognitively disabled patients those who have the capacity to care or to maintain simple relationships and those who do not. Those who have such capacities clearly benefit from life, whereas those who lack even these simple capacities neither benefit from nor are harmed by continued life.
26. Despite such reports, there remains some controversy about what the life of a locked-in patient amounts to, with many arguing that such a life must be one of unbearable suffering. Indeed, because many people think such lives must be awful, locked-in patients are sometimes allowed to choose death relatively soon after their diagnosis—which would not happen, for example, with many other severe disabilities, where it would be assumed that patients need time to adjust. (See, for example, Kompanje, de Beaufort, & Bakker, 2007.) I allow that for those rare patients who are *completely* locked-in (i.e., cannot communicate even via eye signals), continued life may simply be a continuation of suffering. But this is based on the same reasoning as my claim that minimally conscious patients, who are nonetheless somewhat cognitively intact, almost certainly suffer. In both cases, relationships are impossible. This is also true for those patients who have not been recognized by others as being locked in, but at least here there is hope for improvement. Were someone to recognize their state, communication could be established. The locked-in patient Julia Tavalaro wrote a memoir describing (among other things) her frustration and misery during the 6 years it took for others to realize she was locked-in (Tavalaro & Tayson, 1997). All of this is compatible

with the thought that in most cases being locked-in need not lead to unbearable suffering, because in most cases communication remains possible and, therefore, so do relationships.

27. The most common methods make use of the fact that the patient can blink or move the eye up and down. Using a yes/no system and alphabet listings, patients can select one letter at a time, thus creating their own communications that are not dependent on the questions of others (see Laureys et al., 2005, pp. 501–503, 505–506).

REFERENCES

Albrecht, G. L., & Devlieger, P. J. (1999). The disability paradox: High quality of life against all odds. *Social Science and Medicine, 48*, 977–988.

Arras, J. (1984). Toward an ethic of ambiguity. *The Hastings Center Report, 14*(2), 25–33.

Borthwick, C. J. (1996). The permanent vegetative state: Ethical crux, medical fiction? *Issues in Law and Medicine, 12*, 167–185.

Brickman, P., Coates, D., & Janoff-Bulman, R. (1978). Lottery winners and accident victims: Is happiness relative? *Journal of Personality and Social Psychology, 36*(8), 917–927.

Buchanan, A., & Brock, D. (1990). *Deciding for others: The ethics of surrogate decision-making.* New York: Cambridge University Press.

Cassell, E. J. (1982). The nature of suffering and the goals of medicine. *New England Journal of Medicine, 306*(11), 639–645.

Cassell, E. J. (1991). Recognizing suffering. *The Hastings Center Report, 21*(3), 24–31.

Cruse, D., Chennu, S., Chatelle, C., Bekinschtein, T. A., Fernández-Espejo, D., Pickard, J. D., . . . & Owen, A. M. (2011). Bedside detection of awareness in the vegetative state: A cohort study. *Lancet, 378*, 2088–2094.

Damschroder, L., Zikmund-Fisher, B. J., & Ubel, P. (2008). Considering adaptation in preference elicitations. *Health Psychology, 27*(3), 394–399.

Doble, J. E., Haig, A. J., Anderson, C., & Katz, B. (2003). Impairment, activity, participation, life satisfaction, and survival in persons with locked-in syndrome for over a decade: Follow-up on a previously reported cohort. *Journal of Head Trauma Rehabilitation, 18*(5), 435–444.

Dresser, R. S., & Robertson, J. A. (1989). Quality of life and non-treatment decisions for incompetent patients: A critique of the orthodox approach. *Law, Medicine, and Healthcare, 17*(3), 234–244.

Frankfurt, H. (1989). On caring. In H. Frankfurt. *Necessity, volition, and love.* Cambridge: Cambridge University Press.

Giacino, J. T., Ashwal, S., Childs, N., Cranford, R., Jennett, B., Katz, D. I., . . . & Zasler, N. D. (2002). The minimally conscious state: Definition and diagnostic criteria. *Neurology, 58*, 349–353.

Griffin, J. (1986). *Well-being: Its meaning, measurement, and moral importance.* Oxford: Oxford University Press.

Jaworska, A. (1999). Respecting the margins of agency: Alzheimer's patients and the capacity to value. *Philosophy and Public Affairs, 28*, 105–138.

Jaworska, A. (2007). Caring and internality. *Philosophy and Phenomenological Research, 64*(3), 529–568.

Kahneman, D. (2011). *Thinking, fast and slow.* New York: Farrar, Straus, and Giroux.
Kim, S. (2009). *Evaluation of capacity to consent to treatment and research.* New York: Oxford University Press.
Kompanje, E. J. O., de Beaufort, I. D., & Bakker, J. (2007). Euthanasia in intensive care: A 56-year old man with a pontine hemorrhage resulting in a locked-in syndrome. *Critical Care Medicine, 35*(10), 2428–2430.
Laureys, S., Pellas, F., Van Eeckhout, P. (2005). What is it like to be conscious but paralyzed and voiceless? *Progress in Brain Research, 150,* 495–511.
Matter of Conroy. (1985). 486 A.2d 1209, 98 N.J. 321, 98 New Jersey 321.
McCormick, R. A. (1990). To save or let die. In T. A. Shannon & J. J. Walter (Eds.). *Quality of life: The new medical dilemma* (pp. 26–34). Mahwah, NJ: Paulist Press.
Monti, M. M., Vanhaudenhuyse, A., Coleman, M. R., Boly, M., Pickard, J. D., Tshibanda, L., . . . & Laureys, S. (2010). Willful modulation of brain activity in disorders of consciousness. *New England Journal of Medicine, 362*(7), 579–589.
Multi-Society Task Force on Persistent Vegetative States. (1994). Medical aspects of the persistent vegetative state. *New England Journal of Medicine, 330*(21), 1499–1508.
Pence, G. E. (2008). *Classic cases in medical ethics.* 5th ed. Boston, MA: McGraw-Hill.
Robertson, J. A. (1975). Involuntary euthanasia of defective newborns: A legal analysis. *Stanford Law Review, 27,* 213–269.
Sinnott-Armstrong, W. (2011, December 9). *Is he conscious, does he want to be?* Humanities in Medicine Lecture. Durham, NC: Duke University Medical Center, Trent Center for Bioethics, Humanities & History of Medicine.
Steffen, G. E., & Franklin, C. (1985). Case studies: Who speaks for the patient with locked-in syndrome? *Hastings Center Report, 15*(6), 13–15.
Stone, J. (2007). Pascal's wager and the persistent vegetative state. *Bioethics, 21*(2), 84–92.
Tavalaro, J., & Tayson, R. (1997). *Look up for yes.* New York: Kodansha America.

12

Minimally Conscious States and Pain

A Different Approach to Patient Ethics

VALERIE GRAY HARDCASTLE

If we were honest, we would have to confess that we do not know much about the neural components of consciousness. Some neurologists are arguing, however, that new studies of minimally conscious and vegetative patients using functional magnetic resonance imaging (fMRI) give us important clues to the neural circuitry required for consciousness. At the same time, other fMRI studies highlight areas of the brain that are activated by noxious stimuli in these patients. Some are putting these two threads of research together to conclude that vegetative patients cannot experience pain but minimally conscious patients probably do. In this essay, I shall suggest that both strands of argument are faulty. In particular, I claim that understanding conscious experiences is not directly relevant for determining whether someone is in pain, and the type of disorder of consciousness (DOC) one has is relevant neither to being in pain nor to how pain should be treated.

STATING THE OBVIOUS

I shall begin by stating what I believe to be obvious: We do not know what consciousness is. A quick scan of the literature (or the Internet) tells us that definitions of consciousness abound. Following John Locke's (1688) original conception of consciousness as "the perception of what passes in a man's own mind," contemporary philosophers in the main think of consciousness in terms of subjective experience, either of the world or of oneself. Consciousness is "what it is like" to be. This is a virtually ineffable instance of having a particular first-person point of view, a perspective made most famous by Thomas Nagel (1974).

Although these approaches suffice for philosophers, they do not serve other intellectual communities as well. Scientists and doctors, for example, would prefer to operationalize the concept so that they could test empirically whether someone or something has consciousness. As a result, many different cognitive functions have been proposed as the index for consciousness. Some have asserted that consciousness is co-extensive with attentional processes (e.g., Graziano & Kastner, 2011; Posner, 1994). Others have suggested linguistic ability, means-ends reasoning, episodic memory, or some permutation of these attributes (e.g., Macphail, 1998, Tononi & Laureys, 2009). Perhaps more important for our purposes, the Multi-Society Task Force on Persistent Vegetative States specifically required self-awareness for consciousness (1994), although they have been criticized rather extensively for being so narrow in their definition, and the American Medical Association (1990) declared that consciousness has two components—arousal and cognitive content—which underlie or support learning, memory, self-awareness, and other functions.

Neuroscientists have taken a different approach and have focused on uncovering so-called neural correlates of consciousness, the essential substrates or processes for consciousness to exist. Since this project's inception in the early 1990s with Francis Crick's and Christof Koch's (1990) discussion of the binding problem, the search for the neural correlates of consciousness has gained both popularity and respectability. However, it does not appear that we are homing onto any answers. Recent suggestions for what these correlates are include global integrated fields (Kinsbourne, 1988), synchronous oscillation (Crick & Koch, 1990; Singer, 1999), NMDA-mediated transient assemblies (Flohr, 1995), thalamocortical patterns of activation (Llinas, 2001), reentrant cortical loops (Edelman, 1989), action-prediction-assessment comparator mechanisms in frontal and midbrain areas (Gray, 1995), affective somatosensory hemostatic processes based in the frontal-limbic nexus (Damasio, 1999), left hemisphere–based interpretative processes (Gazzaniga, 1988)—and the list goes on. Recent speculations also include Giulio Tononi's suggestion (Tononi, 2004; Tononi & Edelman, 1998) that consciousness is related to integrating information from disparate regions of the brain into a coherent whole, which dovetails well with recent research in disorders of consciousness and understanding the difference between persistent vegetative states (VS) and minimally conscious states (MCS) (see later discussion).

States have gotten into the act as they craft laws and policies that govern surrogate end-of-life decision making. Some, like Florida, Alabama, Delaware, Connecticut, and Maryland, define consciousness behaviorally: Consciousness refers to how one interacts with the environment or communicates with others.

Others, such as Idaho, Maine, Ohio, and West Virginia, define consciousness functionally: Consciousness becomes identified with a cognitive or brain trait. Still others, such as Kentucky, use a hybrid notion. Courts, too, have oscillated between behavioral and functional definitions. In the 1976 seminal decision involving Karen Ann Quinlan, the courts relied on a functional understanding: A vegetative person, one who is not conscious, is someone who "no longer has any cognitive function." In contrast, in 1987, the same New Jersey Supreme Court used a behavioral definition in the *Matter of Jobes*: Unconsciousness referred to "no behavioral awareness of the surroundings in a learned manner" (see discussion in Eisenberg, 2008).

Nevertheless, at least some of those who investigate persistent vegetative or minimally conscious patients believe that they can detect conscious awareness in noncommunicative patients by looking at whether and how their brains appear to process cognitive information. Members of Steven Laurey's Coma Science Group hold that "coactivation of specialized sensory cortices and frontoparietal areas seems both necessary and sufficient to generate conscious perception" (Boly et al., 2008, p. 1018). In contrast, if "primary cortical activation . . . appears to be isolated from higher-order associative cortical activity," then the patient is in a persistent vegetative state or has unresponsive wakefulness syndrome (UWS) and is not conscious (Schnakers, Chatelle, Demertzi, Majerus, & Laureys, 2012, p. 438). Martin Monti and Audrey Vanhaudenhuyse believe that fMRI data can provide "clear evidence" that a patient is "aware and able to communicate" (Monti et al., 2010, p. 576). Adrian Owen and his colleagues are quite definitive in their conclusions, asserting that the fMRI patterns they saw "confirmed beyond any doubt that [their patient] was consciously aware of herself and her surroundings" (Owen et al., 2006, p. 1402).

I am claiming that such conclusions and assertions are premature, but more importantly, I am claiming that, relative to questions regarding the ethical treatment of patients, these conclusions and assertions miss the mark entirely.

The immediate challenge is that there really is no consensus over what consciousness is. As Ned Block has remarked: "The concept of consciousness is a hybrid or better, a mongrel concept: the word "consciousness" connotes a number of different concepts and denotes a number of different phenomena" (2002, p. 206). The late Alan Cowey was right when he wrote the following:

> Three common maxims about consciousness are: (1) those who study consciousness agree about only one thing, namely there is no agreement about its definition; (2) consciousness has as many theories as theorists; and (3) consciousness is what brain scientists become interested in towards the end of their careers. (1997, p. 54)

(At least he was right about his first two maxims; the study of consciousness has become much more a field in its own right since the 1990s.) But even as late as the mid-2000s, scientists were still apologizing for their lack of progress: "We have no idea how consciousness emerges from the physical activity of the brain and we do not know whether consciousness can emerge from non-biological systems, such as computers," claimed the authors of one textbook on brain function (Frackowiak et al. 2004, p. 269).

In sum, we do not have a good definition for consciousness, we do not know what the relevant psychological attributes of consciousness are, and we have no idea what the neural correlates for consciousness are either. We are not clear on what is sufficient for consciousness, and we at best have an incomplete list of what is necessary. We do not understand the relationship between alertness and awareness, if there is one, nor do we understand the connection between cognitive processing and consciousness, if there is one. At best, we can point to some things that some people believe index some aspects of consciousness. But by the standards of contemporary science and medicine, that is not pointing to very much at all.

Maybe this should make you suspicious about reading a book on disorders of consciousness and, in particular, reading its chapters on the ethical treatment of such patients. At the very least, it should probably give you pause regarding claims being made by scientists and doctors about whether patients in a persistent vegetative state or those in minimally conscious states are in fact aware of anything.

But the good news amid all this naysaying is that it probably does not matter at all that we do not know one whit about consciousness, who has it, or what it is. At least, it does not matter for this discussion regarding how to understand pain processing in patients with disorders of consciousness. Allow me to explain.

CONSCIOUSNESS AND MORAL SIGNIFICANCE

Why is it that we should care whether others are conscious? Why is discerning a patient's level of awareness important to doctors and the patient's family? One reason we find knowing about the consciousness of others so compelling is that we assign moral worth to creatures based on their putative level of consciousness.

This connection abounds, for example, in moral arguments for vegetarianism (e.g., Singer 1975). After searching for the neural underpinnings of consciousness for most of his professional career, Christoph Koch (2012) publicly converted to vegetarianism because he saw too many similarities between human brains, and therefore human consciousness, and those of other animals. He argued thusly: Other animals are likely conscious, because they resemble us. Therefore, they likely feel pleasure and pain, as well as happiness, sadness, boredom, fear, and so on. Therefore, we should not eat them, just as we should not eat any sentient creature.

Logically speaking, how can Koch get to his final conclusion? He can only if he has as a suppressed premise that conscious creatures have inherent moral worth in virtue of their consciousness. Koch believes that conscious creatures have a special place in the moral hierarchy of our universe; they have intrinsic value. And because of this, we should treat them differently—we should treat them with care and respect. In other words, we should treat them as Kantian ends-in-themselves and not as mere means (such as nourishing us as a meal).

Closer to our concerns in this essay, similar considerations arise in discussions of brain death and abortion. The United Kingdom uses a brainstem definition of death (rather than defining it as the cessation of activity in essentially the whole brain, as the United States does; see Chapter 4 for further discussion). It is argued that loss of activity in the brainstem necessarily means permanent loss of consciousness due to loss of the reticular activating system, among other areas (cf. Kahane & Savulescu, 2009). In the United States, the advisability of late-stage abortions is undergoing renewed scrutiny, largely because questions about whether fetuses can experience pain and, if so, at what point such experiencing begins, remain unanswered (or at least unanswered to everyone's satisfaction).

I am not going to argue against the idea that we should hold being conscious fundamental to imbuing a creature with inherent value; I am just accepting as given that this perspective is widespread in contemporary Western society. Many people (perhaps most people) believe that being conscious elevates one's moral status in the world. My only point here is that this is, in fact, a deeply held premise in most folks' moral reasoning—so deep, in fact, that we do not bother to defend that premise in moral discussions, science policy briefs, or medical texts. Indeed, we rarely even acknowledge its existence.

We can see the way these ideas are fundamentally woven into our cultural fabric in the sad case of Terri Schiavo. The Florida court battle over whether to allow her husband to remove her feeding tube turned on, among other things, whether Schiavo was in a persistent vegetative state or minimally conscious. The assumption all around was that if she were minimally conscious, if she had at least fleeting awareness of her surroundings and the things in them, then her life should be preserved (see Horne, 2009; Kahane & Savulescu, 2009; McCullagh, 2004 for discussion). The unargued assumption was that her life would be worth more if she were just barely conscious. (The courts ruled that Schiavo was in fact in a persistent vegetative state, that her husband had the legal right to determine the course of her medical treatment, and that the legislative and executive branches of the government were blocked from intervening by establishment clause of the First Amendment. Schiavo died 13 days after her feeding tube was removed, and autopsy revealed that her brain had atrophied significantly in the 15 intervening years after the hypoxic event.)

As an interesting side note, Florida law is written so that a surrogate decision maker for a minimally conscious patient can in theory refuse artificial nutrition or hydration, because it defines consciousness behaviorally. Its statutes refer to the "absence of voluntary action or cognitive behavior of any kind" and the "inability to communicate or interact purposefully with the environment." In addition, it authorizes surrogate decision making for an "end-stage condition," which it defines as "an irreversible condition that is caused by injury, disease, or illness which has resulted in progressively severe and permanent deterioration" (cf. Eisenberg, 2008). This language does not exclude many patients with diagnoses of minimally conscious states from falling under its statutory authorization, including perhaps Terry Schiavo. But I digress.

What I am going to focus on is the flip side of the deep assumption that being conscious conveys something special. What we need to examine is whether being conscious is the *only* way to elevate our moral status. Could entities that are not conscious also have inherent value? The founder of the categorical imperatives himself might have thought so, for rationality was the only requirement Immanuel Kant (1785) advanced for being an end-in-itself. According to Kant, being able to choose a course of action to achieve a set of goals, as opposed to merely following instincts and desires, is what sets us humans apart from other animals. And it is this ability that gives us special moral status.

Others connect moral significance with the possession of interests. For example, Kahane and Savulescu wrote that, "it is a truism that interests matter morally" (2009, p. 11). What counts as an interest can vary widely, depending on what sort of ethics you happen to hold dear. Hedonists tie interests to being happy or experiencing pleasure. Deontologists hold that certain activities are intrinsically good, such as developing talents or pursuing knowledge. Desire-satisfaction theorists believe that satisfying our desires is what is important. And so on. For the purposes of this essay, it is not relevant whether any of these views are correct. If one could be happy, engage in certain activities, or have one's desires satisfied without being conscious, then we can conclude that there are ways of understanding moral worth that do not require being conscious.

Not everyone believes that we could do or have these things without being conscious. Hassoun and Kriegel (2008), for example, argue that to be a morally significant being requires not only consciousness but consciousness of self, or something like it. The problem with this sort of attitude, however, is that it creates an artificial divide between conscious and unconscious processing. It implies that conscious cognition, perception, emotion, and thought are important but that unconscious processes such as subliminal perception, cognition, and affect apparently have no significance. This sort of attitude also assumes that consciousness is completely separate and separable from our unconscious states. I submit that these positions are both mistakes.

We have known since the late 1800s that our brains process stimuli presented below the threshold of conscious awareness, because subjects were able to respond correctly to questions regarding the putatively unperceived stimuli above the level of chance (e.g., Sidis, 1898). Classic studies conducted in the 1970s by Anthony Marcel (1974, 1983) further demonstrated that unconsciously or subliminally perceived stimuli influence decisions regarding consciously perceived stimuli. Investigations of memories of unconsciously perceived events, such as perceptions of things that occurred while a patient was under anesthesia, have provided further evidence of subliminal processing (cf. Merikle & Daneman, 1996). Additional evidence for the existence of unconscious processing has come from the study of neurological disorders in which conscious perception of a variety of inputs is specifically impaired but the ability to process and make use of such input remains preserved (e.g., blindsight, deaf hearing, numb touch, the various agnosias) (Barton, 2003; Cowey, 2010; Moscovitch, Winocur, & Behrmann, 1997; Weiskranz, 1996). Many subsequent studies have explored the processing power of unconscious cognition (see Merikle & Daneman, 1998, and Merikle, Smilek, & Eastwood, 2001 for review and discussion).

Currently, we really have no idea what distinguishes conscious from unconscious phenomena in the brain. Consequently, it is likely a mistake to make too much of the distinction between the two, especially from a neurological perspective. In particular, it is likely a mistake, from my vantage point anyway, to make the division between consciousness and unconsciousness what culls morally worthy creatures from the unworthy. That difference is just too shaky and poorly understood to be able to support such an important partition.

PAIN PROCESSING IN PATIENTS WITH DISORDERS OF CONSCIOUSNESS

Doctors and families often want to know whether their patients are conscious because they believe the answer should inform how the patients are treated. The question for many healthcare professionals is whether a patient can feel any pain or is suffering. Their answer to this question determines how much or whether the patient receives analgesia or other sedating medications. Terri Schiavo, for example, did not receive any pain medication after her feeding tubes were disconnected because the courts determined that she was in a persistent vegetative state and, hence, unable to experience pain or discomfort. Indeed, more than half of doctors surveyed in Europe believed that patients in a persistent vegetative or unresponsive wakefulness state do not experience pain (Dermertzi et al., 2009). It is reasonable to expect them to act on these beliefs by not providing analgesic medication for these patients during care or during the dying process after withdrawal of artificial hydration and nutrition (for discussion, see Ahronheim & Gasner, 1990; Fins, 2006).

But is the question of whether a patient is conscious the right one to ask in order to determine level of palliative care? To answer, let us first look at both how pain is defined and how the brain processes it.

Paralleling the definition used by the International Association for the Study of Pain (IASP), the Multi-Society Task Force (1994) decreed that "pain and suffering refer to the unpleasant experiences that occur in response to stimulation of peripheral nociceptive receptors and their peripheral and central afferent pathways or that may emanate endogenously from the depths of human self-perception." In other words, pain is a conscious state caused by nociception and suffering. To determine the level of pain and suffering experienced by a noncommunicative patient, Caroline Schnakers and her colleagues created and validated (and later revised) a Nociception Coma Scale, based on the sorts of scales physicians use to determine the level of pain in infants or in patients with severe dementia (Chatelle et al., 2012; Schnakers et al., 2010, 2012). Table 12.1 outlines the scale, which relies on patient motor, verbal, and visual responses as well as facial expressions. The expectation is that this scale, if correct, measures or indexes the level or depth of pain processing (or of pain experiences) in the brain. (See Chapter 6 for further discussion.)

We know quite a lot (but not everything) about how pain is processed in the brain. In a nutshell, stimulation of peripheral nociceptors activates a dedicated spinothalamic tract that goes to the thalamus and the midbrain, which then transmits pain information up to the cortex. This initial processing is normally referred to as "nociception" and is largely reflexive. Pain processing is more complex; it assigns affective value, behavioral motivation, and cognitive meaning to the stimulus. Historically, this type of processing has been divided into two basic components, an affective-motivational processing pathway and a sensory-discriminatory processing pathway. This basic division goes back to the work of Sir Charles Scott Sherrington and Sir Henry Head on the function of neurons and the somatosensory system in the early 1900s (Sherrington, 1906; Head & Holmes, 1911). Its fundamental importance was underscored in the 1960s as neuroscientists discovered two different spinothalamic tracts involved in nociception, a medial pathway involved in processing affect and a lateral pathway involved in sensory encoding. Neuroscientists

Table 12.1 THE NOCICEPTION COMA SCALE (NCS)

MOTOR RESPONSE
3 — Localization to noxious stimulation
2 — Flexion withdrawal
1 — Abnormal posturing
0 — None/Flaccid
VERBAL RESPONSE
3 — Intelligible verbalization
2 — Vocalization
1 — Groaning
0 — None
VISUAL RESPONSE
3 — Fixation
2 — Eyes movements
1 — Startle
0 — None
FACIAL EXPRESSION
3 — Cry
2 — Grimace
1 — Oral reflexive movement, Startle response
0 — None

(Adapted from Schnakers et al., 2010)

later tied activity in the primary and secondary somatosensory cortices to the sensory-discriminative aspects of pain processing and activity in the cingulate, insula, and prefrontal cortices to the cognitive and affective aspects of pain processing. (In addition to these ascending pathways, there is also a well-defined descending network involving the prefrontal cortex, cingulate, periaqueductal gray, and posterior thalamus that modulates the ascending system. This descending pathway can alter the brain's responses to nociceptive inputs, as well as behavioral and perceptual outputs, such that we see different responses to the same stimuli if the stimuli are given in different contexts.)

However, since the hypothesis for the division between the affective-motivational processing pathway and a sensory-discriminatory processing pathway was put forth in the 1960s, scientists have not made much progress in uncovering the details of its operations. Indeed, as Vania Apkarian maintained, evidence for division between these pathways remains "fairly weak" (2012). He, along with others, advocated a different perspective on pain processing, that it is a subset of our general motivational processing system. The basic idea is that normal acute pain, which signals tissue damage, first prompts us to escape or avoid our current situation in order to minimize physical harm and then, when the pain stops, provides us with a sense of relief. Together, these two reactions—avoidance and relief—not only protect our bodies but also contribute to our being able to predict the utility and costs of competing behavioral goals. That is, when we are confronted with a conflict between a potential threat and a potential reward, we have learned from past pain experience how to evaluate the relative severity of the danger.

A consequence of adopting this perspective on pain processing is that it denies that there is a brain circuit specific to pain processing (Iannetti & Mouraux, 2010). Pain processing is simply part of our ongoing risk-reward calculations. Indeed, it might turn out that Melzack's (1989) original idea for a nonspecific neuromatrix, a widely distributed network of neurons that crosses many areas of the brain and underlies pain perception, might be correct after all. But regardless of which theoretical approach is the correct one, we do know that pain processing involves many different areas of the brain; it is a very complex, widely distributed, and massively parallel process. Imaging studies of healthy subjects experiencing pain bear out the existence of this complex network. They show widespread activation across many cortical and subcortical areas, the details of which need not detain us here. Suffice it to say, they track the data from anatomical studies in rats and other animal models quite well.

What is important for us, though, is the sort of brain damage one typically sees in persistent vegetative and minimally conscious patients and how that damage relates to pain processing in the brain. Research shows that there are large-scale dysfunctions in the frontal regions, the temporal and parietal areas, and the cingulate and precuneal cortices in patients with vegetative state/unresponsive wakefulness syndrome, and less of such dysfunctions in minimally conscious patients (Laureys et al., 1999; Laureys, Owen, & Schiff. 2004). These brain disruptions are correlated with the degree of clinical impairment these patients have; hence, they might be markers for a patient's relative level of overall cognitive deficiency. Important for understanding pain processing in these patients are imaging studies which have shown that connectivity within and between the cingulate and the prefrontal cortex is more disrupted in vegetative patients than in minimally conscious patients (Cauda et al., 2009; Vanhaudenhuyse et al., 2010).

Using positron emission tomography, Steven Laureys examined how the brain reacts to nociceptive stimuli in vegetative state/unresponsive wakefulness syndrome patients compared with healthy controls (Laureys et al., 2002). In all of the patients, only the midbrain, thalamus, and primary somatosensory cortex were activated in response to electrical stimulation of the median nerve, which demonstrates at least intact nociceptive processing and perhaps some higher-level processing as well. Later studies found not only these areas to be active in persistent vegetative patients during pain processing, but also secondary somatosensory cortex, the cingulate, and the posterior insula (Kassubek et al., 2003). However, in no cases were brain activation responses close to what one finds in healthy controls. Vegetative patients' cortical areas appear to be functionally disconnected from one another and from other subcortical areas, and this presumably would lead to incomplete pain processing (Boly et al., 2005). In particular, recent work suggests that vegetative patients lack many of the feedback connections from the association cortices (Boly et al., 2011). Investigators take this evidence to imply that vegetative patients lack the ability to integrate information from different areas of the brain, which they believe is crucial for conscious experience. They conclude that whatever pain processing is going on in the brain of a vegetative state patient, it is likely to be unconscious (cf. discussion in Schnakers et al., 2012).

Indeed, the Multi-Society Task Force concluded in 1994 that the pain behaviors which patients in persistent vegetative states often exhibit—grimacing, posturing, crying, even racing heartbeats and hormonal fluctuations—are not indicative of actual pain or suffering "unless they are consistent, sustained, and definitive in nature." Although nociceptive stimuli can elicit unconscious behavioral responses, as well as other autonomic or endocrinological reflexes, the Task Force stated that this can happen without the patient's having any experience of pain or suffering if the brain has lost its capacity for self-awareness. The imaging data that has accrued since the Task Force report has only supported its findings.

Baseline integrative capacities in minimally conscious patients do appear to be better preserved than those in vegetative state patients (Vanhaudenhuyse et al., 2010). Not surprisingly, we also see brain responses to nociceptive stimuli in minimally conscious patients that are very similar to those in healthy controls, with activation in the thalamus, primary and secondary somatosensory cortices, insula, and cingulate (Boly et al., 2008). Also not surprisingly, investigators have concluded that minimally conscious patients are likely to have the brain integrity necessary for conscious awareness of nociceptive stimuli: "the co-activation of specialized sensory cortices and frontoparietal areas seems both necessary and sufficient to generate conscious perception. . . . We interpret the brain activation and functional connectivity patterns seen in patients in MCS as likely to show conscious perception of noxious stimuli" (Boly et al., 2008, p. 1018; cf. Schnakers et al., 2012). But regardless of whatever the conscious states of minimally conscious patients may be, their neural responses to nociception and pain are virtually identical to those of normal subjects.

I do not wish to adjudicate how the neural systems of patients in a vegetative or minimally conscious state do or do not respond to nociception and pain relative to neural systems in healthy subjects. Nor am I interested particularly in whether one feels pain in any of these conditions. It is sufficient to note that vegetative state and minimally conscious patients alike process nociceptive information, and they both also process pain to some degree. These facts are enough to raise the question regarding how one should treat vegetative state or minimally conscious patients in pain.

We already know that, in the main, doctors and other healthcare providers assume that patients in a vegetative state are not conscious and therefore can neither feel pain nor suffer. Most also assume that minimally conscious patients have some awareness of their bodily conditions and therefore are likely feel pain and could suffer. Consequently, there is no real impetus to provide pain relief to patients in vegetative states.

Even when we look at what the neurologists who study pain processing in patients with disorders of consciousness say about treating these patients, we see similar lines of reasoning. At best, the authors note that upwards of 40% of patients in putative persistent vegetative states are misdiagnosed (cf. Schnakers et al., 2009), so they urge prudence in providing analgesics to vegetative state patients:

> The data suggest that these MCS patients could perceive pain. On the contrary, VS patients showed a functionally disconnected brain activity, suggesting the absence of an integrated pain perception. Considering these results, adequate analgesic treatment has to be provided in MCS patients. The issue is much more complicated in VS patients. Given the high rate of misdiagnosis . . ., if we decide not to administer analgesic treatment in the presence of a potential painful experience (e.g., contractures or fractures), there is a real probability for not treating a patient erroneously diagnosed and, hence, for not treating a patient who perceives pain. (Schnakers et al., 2007)

> Considering the levels of clinical uncertainty, pain treatment should be considered in all patients diagnosed as being in a VS/UWS or MCS. (Schnakers et al., 2012, p. 442)

> It is our opinion, based on both extensive clinical experience and recent data questioning the ability of the bedside exam, even when sophisticated, to delineate awareness in at least some patients with DOCs, that one is likely taking the safer course by treating all DOC patients as if they had the potential to perceive pain and suffer. (Schnakers & Zasler, 2007, p. 624)

I suggest that it is important to know that both persistent vegetative patients and patients in minimally conscious states process nociceptive stimuli and engage in what we would commonly recognize as at least some aspects of pain processing. I submit, however, that it is not important to know which aspects of the processing are conscious. Here is why.

THE IMPACT OF PAIN

In a nutshell, processing of pain, in and of itself, has a negative effect on the brain and body. We gain support for this conclusion through several very different strands of evidence. I will briefly outline three of them.

Preterm infants normally undergo multiple painful experiences, such as heel sticks, intravenous catheters, chest tubes, endotracheal suctioning, and surgery. Indeed, the sickest of premature babies are subjected to an average of 750 procedures before their initial discharge from hospital (Porter et al., 1999). Premature infants in Canadian hospitals normally endure two to eight painful procedures each day. And yet, analgesics are provided in fewer than 10% of the cases (Johnston et al., 1997; Stevens et al., 2003). However, investigations of nociception in neonatal patients indicate that preterm infants are more sensitive to pain than full-term infants and, more importantly, that acute nociceptive stimuli lead to extended episodes of hyperalgesia, during which nonpainful stimuli can induce chronic pain (cf. Anand, 1998; Page, 2004). In addition, nociceptive stimuli in neonates has been linked to early intraventricular hemorrhage and periventricular leukomalacia, a type of white-matter brain injury. Finally, administering analgesics to neonates is correlated with fewer episodes of these types of injuries (Anand, 1998).

And we do not need to focus on preterm infants to see long-term effects of painful stimuli. Coded pain scores during immunization in male babies were linearly related to type of circumcision analgesia (no circumcision, topical agents, or placebo analgesics) (Taddio, Katz, Ilersich, & Koren, 1997). There are also some data linking early pain experiences to permanent changes in spinal cord processing as well as a compromised immune system and even some behavioral disorders such as hyper-vigilance, sleep disturbances, avoidance behaviors, and feeding problems (cf. Mitchell & Boss, 2002; Page, 2004). Regardless of the considerable discussion regarding whether young infants are conscious or, if conscious, whether they can remember painful events, traces of those events linger in the body and the brain, altering their developmental trajectory in fundamental ways.

We can see similar permanent effects with chronic pain. Chronic pain is represented in different areas of the brain than acute pain because, as it turns out, the brain reorganizes itself when it is in chronic pain. Most types of chronic pain shift away from what might be referred to as simple nociceptive processing to more affective and motivational reactions. Moreover, we see specific ongoing non-pain effects; for example, the baseline levels of activity in the insula and anterior cingulate are much higher in chronic pain patients than in control subjects.

In addition, the way in which the nucleus accumbens connects to the rest of the brain is different in healthy subjects and chronic pain patients. In normal subjects, the nucleus accumbens and the insula are tightly connected to each other. But in chronic pain patients, the nucleus accumbens shifts its functional connectivity to medial prefrontal cortex (Baliki, Geha, Fields, & Apkarian, 2010). And the more chronic pain the patient feels, the stronger the correlation between activity in nucleus accumbens and in medial prefrontal cortex. As a result of this rewiring, nucleus accumbens activity differs between healthy subjects and chronic pain patients for instances of acute pain as well. Finally, this change in brain connectivity has to be a functional rewriting not specific to pain processing, because we see no effects for monetary rewards in chronic pain patients—their brains show no real response to either reward or loss (Apkarian 2012). In other words, chronic pain puts stress on our motivational and reward systems such that our systems fundamentally change how they operate. And this change in functionality is so large that it distinguishes between normal subjects and chronic pain patients with an accuracy of more than 90% (Baliki et al., 2011).

These results are even more remarkable because they have been shown to be independent of subjective pain perception. The distorted brain responses are not reflected in the conscious pain ratings by chronic pain patients, and the patients are unaware of any impact their pain sensations have on any aspect of their cognitive processes. The effect of chronic pain on the brain is as an unconscious influence on later actions and decisions.

As several authors discuss in this volume, it is becoming apparent that some patients with severe brain damage can engage in at least minimal decision making, and, of course, we know very little regarding their conscious states—whether they have them at all, or under what circumstances they might have them. The question I am raising is this: Given that we know about the negative impact pain processing has on the brain in a variety of ways, and given the dearth of information we have about what really is going on with patients in a vegetative or minimally conscious state, what sort of pain interventions are appropriate for this patient population?

Pain processing also changes brains morphologically. With long-term or repeated pain, we can measure a significant decrease in gray matter in the areas associated with "neuromatrix" in the brain, especially in the anterior cingulate cortex, right insular cortex, dorsolateral prefrontal cortex, amygdala, and brainstem (cf. Rodriguez-Raecke, Niemeier, Ihle, Ruether, & May, 2009). A few studies have suggested that areas that exhibit the most change in gray matter density also include the hippocampus, multiple lateral frontal regions, and portions of the occipital lobe, suggesting that the morphometric changes are not limited to pain-specific regions (Baliki et al., 2011). Either way, the observed morphological differences in chronic pain conditions correlate with the length of time pain patients have been suffering as well as the intensity of their pain (Baliki et al., 2011).

All of these lines of evidence suggest a new way to think about treating pain. Regardless of level of consciousness in patients, it is clear that processing pain information is bad for their brains and other functional systems. It is also clear that blocking brains from processing pain information prevents these pain-related disruptions. Therefore, it seems clear (to me at least) that whether a patient is conscious has little bearing on whether one should treat that patient's pain. We should treat pain because not to do so is detrimental to the system. All else being equal, minimizing pain activation is always preferable.

The question of how and whether to treat putative pain in patients with disorders of consciousness is a real issue, too, for there are a whole host of reasons why such a patient might be processing pain. In the acute stage of injury, patients often experience fractures, internal injuries, soft tissue injuries, and then the pain associated with invasive surgical procedures. In the chronic stage, patients can exhibit spasticity, contractures, pressure sores, ischemia, peripheral nerve injuries, and postsurgical incisional pain. In addition, headaches are very common after traumatic brain injury, as well as central pain from nerve damage or stroke. It makes a difference how we approach pain processing in patients with disorders of consciousness.

I conclude as follows: How and whether to treat patients with severe disorders of consciousness should not depend on the type of disorder or level of consciousness, alertness, or awareness. It should instead depend on known brain circuitry and neurophysiological and behavioral responses. We should focus on maximizing the good for the system as a whole, regardless of consciousness. Therefore, we should treat pain in all patients, even those we believe are not conscious.

REFERENCES

Ahronheim, J. C., & Gasner, M. R. (1990). The sloganism of starvation. *Lancet*, *335*, 278–279.

American Medication Association. (1990). Persistent vegetative state and the decision to withdraw or withhold life support. *JAMA*, *263*, 426–430.

Anand, K. J. S. (1998). Clinical importance of pain and stress in preterm neonates. *Neonatology*, *73*, 1–9.

Apkarian, A. V. (2012). Chronic pain and addiction pathways. Presented at the Society of Neuroscience Annual Meeting, New Orleans.

Baliki, M. N., Geha, P. Y., Fields, H. L., & Apkarian, A. V. (2010). Predicting value of pain and analgesia: Nucleus accumbens response to noxious stimuli changes in the presence of chronic pain. *Neuron*, *66*, 149–160.

Baliki, M. N., Schnitzer, T. J., Bauer, W. R., & Apkarian, V. (2011). Brain morphological signatures for chronic pain. *PLoS ONE*, *6*, e26010.

Barton, J. J. (2003). Disorders of face perception and recognition. *Neurology Clinic*, *21*, 521–548.

Block, N. (2002). Some concepts of consciousness. In D. Chalmers (Ed.). *Philosophy of mind: Classical and contemporary readings* (pp. 206–218). New York: Oxford University Press.

Boly, M., Faymonville, M.-E., Peigneux, P., Lambermont, B., Damas, F., Luxen, A., . . . & Laureys, S. (2005). Cerebral processing of auditory and noxious stimuli in severely brain injured patients: Differences between VS and MCS. *Neuropsychological Rehabilitation*, *15*, 283–289.

Boly, M., Faymonville, M.-E., Schnakers, C., Peigneux, P., Lambermont, B., Phillips, C., . . . & Laureys, S. (2008). Perception of pain in the minimally conscious state with PET activation: An observational study. *Lancet Neurology*, *7*, 1013–1020.

Boly, M., Garrido, M. I., Gosseries, O., Bruno, M. A., Boveroux, P., Schnakers, C., . . . & Friston, K. (2011). Preserved feedforward but impaired topdown processes in the vegetative state. *Science*, *332*, 858–862.

Cauda, F., Micon, B. M., Sacco, K., Duca, S., D'Agata, F., Geminiani, G., & Canavero, S. (2009). Disrupted intrinsic functional connectivity in the vegetative state. *Journal of Neurology, Neurosurgery, and Psychiatry*, *80*, 429–431.

Chatelle, C., Majerus, S., Whyte, J., Laureys, S., & Schnakers, C. (2012). A sensitive scale to assess nociceptive pain in patients with disorders of consciousness. *Journal of Neurology, Neurosurgery, and Psychiatry*, *83*, 1233–1237.

Cowey, A. (1997). Current awareness: Spotlight on consciousness. *Developmental Medicine and Child Neurology*, *39*, 54–62.

Cowey, A. (2010). The blindsight saga. *Experimental Brain Research*, *1*, 3–24.

Crick, F., & Koch, C. (1990). Toward a neurobiological theory of consciousness. *Seminars in Neuroscience*, *2*, 263–275.

Damasio, A. (1999). *The feeling of what happens: Body and emotion in the making of consciousness*. New York: Harcourt.

Demertzi, A., Schnakers, C., Ledoux, D., Chatelle, C., Bruno, M.-A., Vanhaudenhuyse, A., . . . & Laureys, S. (2009). Different beliefs about pain perception in the vegetative and minimally conscious states: A European survey of medical and paramedical professionals. *Progress in Brain Research*, *177*, 329–339.

Edelman, G. (1989). *The remembered present: A biological theory of consciousness*. New York: Basic Books.

Eisenberg, J. B. (2008). Schiavo on the cutting edge: Functional brain imaging and its impact on surrogate end-of-life decision-making. *Neuroethics*, *1*, 75–83.

Fins, J. J. (2006). Affirming the right to care, preserving the right to die: Disorders of consciousness and neuroethics after Schiavo. *Palliative Support Care*, *4*, 169–178.

Flohr, H. (1995). An information processing theory of anesthesia. *Neuropsychologia, 33*(9), 1169–1180.

Frackowiak, R. S. J., Ashburner, J. T., Penny, W. D., Zeki, S., Friston, K. J., Frith, C. D., . . . & Price, C. J. (2004). *Human brain function*. 2nd ed. San Diego, CA: Academic Press.

Gazzaniga, M. (1988). *Mind matters: How mind and brain interact to create our conscious lives*. Boston: Houghton Mifflin.

Gray, J. (1995). The contents of consciousness: A neuropsychological conjecture. *Behavior and Brain Sciences, 18*(4), 659–722.

Graziano, M. S. A., & Kastner, S. (2011). Human consciousness and its relationship to social neuroscience: A novel hypothesis. *Cognitive Neuroscience, 2*, 98–113.

Hassoun, N., & Kriegel, U. (2008). Consciousness and the moral permissibility of infanticide. *Journal of Applied Philosophy, 25*(1), 45–55.

Head, H., & Holmes, G. (1911). Sensory disturbances from cerebral lesions. *Brain, 34*, 102–254.

Horne, M. (2009). Are people in a persistent vegetative state conscious? *Monash Bioethics Review, 28*(2), 12.1–12.12.

Iannetti, G. D., & Mouraux, A. (2010). From the neuromatrix to the pain matrix (and back). *Experimental Brain Research, 205*, 1–12.

Johnston, C. C., Collinge, J. M, Henderson, S. J., & Anand, K. J. (1997). A cross-sectional survey of pain and pharmacological analgesia in Canadian neonatal intensive care units. *Clinical Journal of Pain, 13*, 308–312.

Kahane, G., & Savulescu, J. (2009). Brain damage and the moral significance of consciousness. *Journal of Medicine and Philosophy, 34*, 6–26.

Kant, I. (1785/1993). *Grounding for the metaphysics of morals*. 3rd ed. (J. W. Ellington, Trans.). New York: Hackett.

Kassubek, J., Juengling, F. D., Els, T., Spreer, J., Herpers, M., Krause, T., . . . & Lücking, C. H. (2003). Activation of a residual cortical network during painful stimulation in long-term postanoxic vegetative state: A 15O-H2O PET study. *Journal of the Neurological Sciences, 212*, 85–91.

Kinsbourne, M. (1988). Integrated field theory of consciousness. In A. Marcel & E. Bisiach (Eds.). *Consciousness in contemporary science*. Oxford: Oxford University Press.

Koch, C. (2012). *Consciousness: Confessions of a romantic reductionist*. Cambridge, MA: MIT Press.

Laureys, S., Lemaire, C., Maquet, P., Phillips, C., & Franck, G. (1999). Cerebral metabolism during vegetative state and after recovery to consciousness. *Journal of Neurology, Neurosurgery, and Psychiatry, 67*, 121.

Laureys, S., Owen, A. M., & Schiff, N. D. (2004). Brain function in coma, vegetative state, and related disorders. *Lancet Neurology, 3*, 537–546.

Laureys, S., Faymonville, M.-E., Peigneux, P., Damas, P., Lambermont, B., Del Fiore, G., . . . & Maquet, P. (2002). Cortical processing of noxious somatosensory stimuli in the persistent vegetative state. *NeuroImage, 17*, 732–741.

Llinas, R. (2001). *I of the vortex: From neurons to self*. Cambridge, MA: MIT Press.

Locke, J. (1688/1959). *An essay on human understanding*. New York: Dover.

Macphail, E. M. (1998). *The evolution of consciousness*. New York: Oxford University Press.

Marcel, A. J. (1974). Perception with and without awareness. Paper presented at the Experimental Psychology Society Annual Meeting. Stirling, Scotland.

Marcel, A. J. (1983). Conscious and unconscious perception: Experiments on visual masking and word recognition. *Cognitive Psychology, 15*, 197–237.

McCullagh, P. (2004). *Conscious in a vegetative state? A critique of the PVS concept.* New York: Kluwer Academic Publishers.

Melzack, R. (1989). Phantom limbs, the self and the brain. *Canadian Psychology, 30*, 1–16.

Merikle, P. M., & Daneman, M. (1996). Memory for unconsciously perceived events: Evidence from anesthetized patients. *Consciousness and Cognition, 5*, 525–541.

Merikle, P. M., & Daneman, M. (1998). Psychological investigations of unconscious perception. *Journal of Consciousness Studies, 5*, 5–18.

Merikle, P. M., Smilek, D., & Eastwood, J. D. (2001). Perception without awareness: Perspectives from cognitive psychology. *Cognition, 79*, 115–134.

Mitchell, A., & Boss, B. J. (2002). Adverse effects of pain on the nervous systems of newborns and young children: A review of the literature. *The Journal of Neuroscience Nursing, 34*, 228–236.

Monti, M. M., Vanhaudenhuyse, A., Coleman, M. R., Boly, M., Pickard, J. D., Tshibanda, L., . . . & Laureys, S. (2010). Willful modulation of brain activity in disorders of consciousness. *New England Journal of Medicine, 362*, 579–589.

Moscovitch, M., Winocur, G., & Behrmann, M. (1997). What is so special about face recognition? Nineteen experiments on a person with object agnosia and dyslexia but normal face perception. *Cognitive Neuroscience, 9*, 555–604.

Multi-Society Task Force on Persistent Vegetative States. (1994). Medical aspects of the persistent vegetative state. *New England Journal of Medicine, 330*(21), 1499–1508.

Nagel, T. (1974). What is it like to be a bat? *Philosophical Review, 83*, 435–456.

Owen, A. M., Coleman, M. R., Boly, M., Davis, M. H., Laureys, S., & Pickard, J. D. (2006). Detecting awareness in the vegetative state. *Science, 313*, 1402.

Page, G. G. (2004). Are there long-term consequences of pain in newborn or very young infants? *The Journal of Perinatal Education, 13*, 10–17.

Porter, F. L, Wolf, C. M, & Miller, J. P. (1999). Procedural pain in newborn infants: The influence of intensity and development. *Pediatrics, 104*, e13.

Posner, M. I. (1994). Attention: The mechanisms of consciousness. *Proceedings of the National Academy of Science U S A, 91*, 7398–7403.

Rodriguez-Raecke, R., Niemeier, A., Ihle, K., Ruether, W., & May, A. (2009). Brain gray matter decrease in chronic pain is the consequence and not the cause of pain. *The Journal of Neuroscience, 29, 13*746–13750.

Schnakers, C., Chatelle, C., Demertzi, A., Majerus, S., & Laureys, S. (2012). What about pain in disorders of consciousness? *The AAPS Journal, 14*, 437–444.

Schnakers, C., Chatelle, C., Vanhaudenhuyse, A., Majerus, S., Ledoux, D., Boly, M., . . . & Laureys, S. (2010). The Nociception Coma Scale: A new tool to assess nociception in disorders of consciousness. *Pain, 148*, 215–219.

Schnakers, C., Faymonville, M.-E., & Laureys, S. (2007). Ethical implications: Pain, coma, and related disorders. In *Encyclopedia of Consciousness.* Retrieved from http://www.credoreference.com/entry/estcon/ethical_implications_pain_coma_and_related_disorders

Schnakers, C., Vanhaudenhuyse, A., Giacino, J., Ventura, M., Boly, M., Majerus, S., . . . & Laureys, S. (2009). Diagnostic accuracy of the vegetative and minimally conscious state: Clinical consensus versus standardized neurobehavioral assessment. *BMC Neurology, 9*, 35.

Schnakers, C., & Zasler, N. D. (2007). Pain assessment and management in disorders of consciousness. *Current Opinion in Neurology, 20*, 620–626.

Sherrington, C. S. (1906). *The integrative action of the nervous system*. New Haven, CT: Yale University Press.

Sidis, B. (1989). *The psychology of suggestion*. New York: D. Appleton.

Singer, P. (1975). *Animal liberation*. New York: Harper Perennial Modern Classics.

Singer, W. (1999). Neuronal synchrony: A versatile code for the definition of relations. *Neuron, 24*, 49–65.

Stevens, B., McGrath, P., Gibbins, S., Beyene, J., Breau, L., Camfield, C., . . . & Yamada, J. (2003). Procedural pain in newborns at risk for neurologic impairment. *Pain, 105*, 27–35.

Taddio, A., Katz, J., Ilersich, A. L., & Koren, G. (1997). Effect of neonatal circumcision on pain response during subsequent routine vaccination. *Lancet, 349*, 599–603.

Tononi, G. (2004). An information integration theory of consciousness. *BMC Neuroscience, 5*, 42.

Tononi, G., & Edelman, G. (1998). Consciousness and complexity. *Science, 282*, 1846–1851.

Tononi, G., & Laureys, S. (2009). The neurology of consciousness: An overview. In S. Laureys & G. Tononi (Eds.). *The neurology of consciousness: Cognitive neuroscience and neuropathology* (pp. 375–412). New York: Academic Press.

Vanhaudenhuyse, A., Noirhomme, Q., Tshibanda, L. J., Bruno, M. A., Boveroux, P., Schnakers, C., . . . & Boly, M. (2010). Default network connectivity reflects the level of consciousness in non-communicative brain damaged patients. *Brain, 133*, 161–171.

Weiskranz, L. (1996). Blindsight revisited. *Current Opinion in Neurobiology, 6*, 215–220.

PART IV

Law

13

The Legal Circle of Life

NITA A. FARAHANY AND RACHEL ZACHARIAS

Would the law be better served by defining the circle of legal life as the beginning and end of consciousness, rather than using a medical standard of death? Do the legal standards of life and death serve their contextual purpose in law? In this chapter, we explore these questions.

It was once simple to define death, in life and in law, as the moment the heart stopped beating and the lungs stopped moving air into and out of the body. But technological innovation unraveled this simplicity by introducing life support machines that move breath into and out of the body long after an individual ceases to do so autonomously. Although many lives have been saved as a result, these innovations have also introduced legal quandaries, because the moment of death has import for organ donation; for criminal, estate, tort law; and more. Over time, in medicine and in law, brain death replaced the traditional cardiopulmonary standard as the new paradigmatic model of death (Ad Hoc Committee, 1968; *Uniform Determination of Death Act,* 1980). But brain death, too, has proved difficult to universally define, particularly because parts of the brain may persist after other parts die.

Now further challenging an adequate definition of the end of a legal life is recent cutting-edge research finding conscious awareness in some individuals who were previously thought to lack all higher brain capacity. These findings have spurred significant ethical and social debates about technological limitations in identifying the loss of consciousness and life. The potential misclassification of individuals also invites a renewed examination of medical and legal standards of death. Although science alone cannot answer the layered and cultural questions of how we define personhood, life, and death, novel scientific discoveries may guide us to better answers. Perhaps, we posit here, cutting-edge research reminds us that legal standards should not turn solely on current and future technological capacities. Law can and should draw from science and technology to inform when death has occurred, but only insofar as it informs a legal standard for what death means in law.

This essay explores defining legal death as building on, but not determined by, a technologically based concept. And it considers the corollary of defining when legal rights end—namely, when legal rights begin.

BRAIN DEATH

Brain death has a complicated and controversial history. The death of Theresa "Terri" Schiavo in 2005 brought the issue to the forefront of national attention. At 26 years of age, Terri Schiavo collapsed from a full cardiac arrest (Fine, 2005; St. Petersburg Police Department, 1990), possibly due to a potassium imbalance brought on by her history of bulimia. (The cause of collapse was never confirmed (Goodnough, 2005).) Physicians found a heartbeat but diagnosed Schiavo with hypoxic encephalopathy: brain injury from an extended lack of oxygen, severe enough to place her in a coma. Over time, Schiavo regained the ability to breathe autonomously, such that she was diagnosed as being in a *vegetative state* (more recently defined clinically as unresponsive wakefulness syndrome, UWS), which constitutes continued neuronal firing and synapsing in the brainstem (the location of vegetative functions) but no electrical activity whatsoever in the higher-functioning regions of the brain (Fine, 2005). Because Schiavo's brainstem was still active, to an observer, and especially to her family, Terri seemed "awake" and conscious at times. She exhibited circadian sleep-wake cycles, opened her eyes sometimes, and even breathed on her own. Despite this, Schiavo's cerebral cortex, often thought to have a vital role in conscious awareness (Merker, 2007), perception, and response to stimuli, was completely inactive; at autopsy, it was found to have atrophied and weighed half its expected size (Goodnough, 2005). The public debates that ensued considered whether having neurons firing in the brainstem is enough to be considered alive—or if something more is required.

Schiavo's case dominated the national headlines and media as her husband (her guardian *ad litem*) waged a legal battle against her parents about her medical care. Schiavo's husband believed that she had died and that the artificial measures being used to keep her bodily functions active, such as feeding and hydration, were unwarranted. Schiavo's parents were convinced she was still alive because she appeared to be awake and breathing. They maintained that removing life-sustaining feeding and hydration would end her life and would be an illegal act of homicide. At stake and debated across the nation were the biological, moral, and legal standards of when a human life ends. After years of legal battles, Schiavo's husband prevailed, and Schiavo's artificial hydration and nutrition were removed. Schiavo's heart, lung, and other organ functions stopped soon after. Under every concept of the word, Schiavo was dead.

Terri Schiavo's case forced a worldwide conversation about when life ends. Are we alive if we are breathing and can open our eyes? Can concepts as rich and as complicated as life and death be defined by particular bodily functions, or do these words carry deeper meaning, such as having an awareness of oneself and the surrounding world? In other words, does life begin, and end, with consciousness? And do legal standards of death and life necessarily coincide with philosophical ones?

LEGAL DEATH

Before we can tackle these questions, we need to understand how death is defined in legal contexts. How does the American legal system grapple with the challenge of defining life and death, especially with some of the implications of advancements in technology that we have hinted at? Before the 1960s, the legal standard of death was simple: "the permanent cessation of cardiopulmonary function" (DeGrazia, 2005). Medicine at this time was such that once a patient's heart or lungs stopped functioning, there was no way to resuscitate them, and thus the cessation of these organs indicated certain death. The diagnosis was simply made by checking a patient's pulse or observing whether moisture appeared on a mirror held in front of the mouth (DeGrazia, 2005).

In the 1960s, advanced emergency rooms and intensive care units opened, equipped with respirators and defibrillators as well as new medicines capable of stimulating some bodily functions (DeGrazia, 2005; Laureys, 2005). These "extraordinary means of life support" (Uniform Determination of Death Act, 1980) suddenly opened the possibility for a patient's respiration and circulation to be maintained artificially, beyond the point that the body could do so autonomously (DeGrazia, 2005). In other words, these technologies allowed patients who had suffered complete loss of brain function to continue to "live" under the traditional cardiopulmonary standard.

In 1968, the Ad Hoc Committee of Harvard Medical School published a new set of criteria for defining death in the *Journal of the American Medical Association* (Ad Hoc Committee, 1968). Under Dr. Henry Beecher, the committee proposed that patients who had permanently lost all brain function could be diagnosed as "dead" (DeGrazia, 2005). This standard came to be known as the *whole-brain standard* of death. It implied that complete brain function, from cerebral executive functions implicating consciousness to reflexive brainstem functioning, was necessary to determine death this way. The whole-brain standard, distinctly different from the cardiopulmonary standard, defines death specifically as the "irreversible cessation of all functions of the entire brain, including the brainstem" (Ad Hoc Committee, 1968).

Throughout the 1970s, most American states legally adopted wholesale the whole-brain standard of death—a medical-biological definition of death—either alongside the cardiopulmonary standard or as a new, distinct standard (DeGrazia, 2005). But the legal standard of death was not uniform nationwide, so how death came to be defined in law varied across state lines. In response, groups such as the American Bar Association and American Medical Association attempted to propose broad statutes defining death, brought to light by criminal charges for removing life support (Bass, 1979). In 1978, the Uniform Law Commissioners (ULC) created the Uniform Brain Death Act, citing the brain as the most central organ in human functioning and thus deciding that its complete and irreversible cessation should legally imply death (Uniform Determination of Death Act, 1980). After some confusion regarding the cardiopulmonary standard, which the ULC notably left out of their 1978 publication, the statute was updated to the Uniform Determination of Death Act (UDDA) in 1981, which reads as follows (DeGrazia, 2005):

> [A]n individual who has sustained either (1) irreversible cessation of circulatory and respiratory functions, or (2) irreversible cessation of all functions of the entire brain, including the brain stem, is dead. A determination of death must be made in accordance with accepted medical standards.

Importantly, the statute was not titled the Definition of Death Act but the *Determination of Death* Act; it was merely intended to provide for a diagnosis of death, and not a philosophical or cultural definition of death. In addition, by not including specific diagnostic tools, the ULC sought to ensure that the statute would remain dynamic and flexible across individual cases (Uniform Determination of Death Act, 1980).

Currently, whole-brain death is at least a portion of the legal standard of death in every jurisdiction in the United States (Olick, Braun, & Potash, 2009). New York and New Jersey both require an apnea test to confirm the lack of respiration as well; in addition, both offer a "religion clause," citing that "societ[al] interests in uniformity should sometimes yield to and *reasonably accommodate* a patient's religious or moral objection to being determined dead on the basis of neurological criteria" (*New Jersey Advance Directives*, 1991; New York State Task Force, 1986; Olick et al., 2009). Despite interstate agreement, the United States still does not have a strong legal consensus about the legal standard of death. Many still argue that whole-brain death is too narrow or too broad a definition to serve the legal purpose of detaching legal standing and rights. For instance, some argue that this definition allows "killing" live people for the sake of organ donation, whereas others think it keeps people attached to life support who are truly no longer living.

Partly fueling the debate is the distinction between whole-brain death and the *brainstem standard* of death. Adopted legally by the United Kingdom after it was proposed by the Honorary Secretary of the Conference of Medical Royal College in 1976, the brainstem definition claims that death is marked by the "complete and irreversible loss of brainstem function" (Laureys, 2005). The brainstem directs vegetative functioning, such as respiration, wakefulness, blinking, and reflexive movements. As discussed more fully later, proponents of the brainstem standard opined that individuals lacking wakefulness (and, consequently, consciousness) or the ability to breathe are simply not alive.

In reality, there is no large debate between the whole-brain and brainstem standards. Both cases distinguish the time of death as the time at which the "permanent cessation of the critical functions of the organism as a whole" occurs (Laureys, 2005). The whole-brain standard is merely a "safer" standard, possibly fostered by the ancient fear of a misdiagnosis of death.

The real conceptual distinction lies between these standards and the neocortical definition of death, which is defined as the "irreversible loss of the capacity for consciousness and social interaction" (Laureys, 2005). From a biological sense, this is most frequently tied with the "higher brain" standard. Widely championed by Robert Veatch in the 1970s, this standard states that an individual's awareness and self-perception constitute life as opposed to death and that, therefore, only the irreversible cessation of the cortex and front brain is necessary for a diagnosis (Veatch, 1975). In part because of a lack of consensus about what consciousness means, and how and whether it can be measured, this is a definition opposed by many in the medical field. Moreover, one can lose higher brain function but still breathe autonomously, and few would comfortably regard that subset of individuals as having "died." And yet, from a legal perspective about when rights attach and detach to an individual based on their present and expected future capacities, different standards may have different merits based on the contextual legal decisions at issue.

With all of this confusion, why is a legal standard necessary? Couldn't we make decisions about death based on personal instincts and opinions? Most basically, criminal law requires a legal standard of death to define murder (Uniform Determination of Death Act, 1980). How could we otherwise distinguish homicide or attempted murder from assault without a clear standard for when a person has died? If a victim suffers brain death but can breathe with respirators, a cardiopulmonary standard of death would change the crime from homicide to assault. Both tort and estate law require legal standards of death (Uniform Determination of Death Act, 1980). In tort law, a legal standard of death is essential for wrongful death cases and survivors' rights

(Uniform Determination of Death Act, 1980). And the property in an estate cannot be resolved and distributed until an individual has been pronounced legally dead.

In the absence of a clear governing legal standard of death, medicine would be stymied. The act of removing life support from a patient by a physician would become a chargeable homicide. And organ donation would become virtually impossible (DeGrazia, 2005). Organs are viable for transplant only for a short time, and they are much more useful when the body has not been sustained artificially for long periods. Under some brain death definitions, organs can be donated once a patient's brain function ceases, even as the body is sustained on life support, which keeps the organs viable for donation. However, in the case of whole-brain death, doctors harvesting organs for transplantation from individuals with slight amounts of brain activity anywhere, even if not functional, are technically committing homicide. To this day, the Canadian legal system juggles the brain and cardiopulmonary death standards; their national standard still allows both standards to independently determine death. As a result, the medical guidelines for declaring death (including in the cases of organ donation) vary across provinces (Sisler, 2014).

Despite its importance, the legal standard is ripe for revision. Instead of choosing between seemingly conflicting definitions and constantly shifting ones in law, we argue that law should adopt a legal standard of death commensurate with the purpose the standard serves. And we offer one to provoke democratic deliberation on a legal standard of death: marking legal death as the permanent loss of consciousness (Bernat, Culver, & Gert, 1981).

FROM BRAIN DEATH TO CONSCIOUSNESS AWARENESS

While the neocortical and often the higher-brain definitions clearly attribute a permanent loss of consciousness to the indication of death, the other two mainstream definitions of death are subtly, yet fundamentally based on a loss of consciousness. The brainstem standard of death reasons that an individual who is incapable of breathing and wakefulness is no longer living, because these are the two fundamental functions that are lost with cessation of activity in the brainstem. In this case, the two functions go hand in hand: When the brainstem is inactive, neither autonomous respiration nor wakefulness is possible. The whole-brain death is functionally the same: Once the brainstem is dead, there is no capacity for respiration or wakefulness, and the absolute lack of higher brain activity only emphasizes a lack of consciousness.

Could irreversible loss of consciousness provide a conceptually more robust legal standard of death? To realize one's legal rights, protections, and their implications one must have legal capacity. A legal standard should be adaptable to changing science and medicine but not exclusively bound to it. One possible solution in law lies at the foundation of all three medical approaches to death. A standard that cannot be produced or maintained artificially is the irreversible cessation of consciousness. If one lacks consciousness, one cannot sense or perceive oneself or one's environment and cannot have any semblance of an experience of life. When an individual's ability for consciousness ends irreversibly, legal rights based on decision-making capabilities could detach. Of course, it is quite odd or uncomfortable to declare individuals such as Terri Schiavo legally "dead" if they are unaware but can breathe and exhibit levels of wakefulness. And yet, legal rights could cease when an individual loses all present and future potential to perceive himself or herself as an individual or his or her environment. Confusion alone would not suffice here to constitute loss of perception—irreversible cessation of consciousness entails an inability to perceive at all, not a flawed perception of one's surroundings. A legal end to life may arrive with such irreversible loss of consciousness.

EXPLAINING CONSCIOUSNESS

Consciousness has two prongs: wakefulness and awareness.[1] Wakefulness is simply defined by the individual's being awake, characterized by the eyes being open and often with some motor arousal. Wakefulness is determined by circadian sleep-wake cycles, a vegetative function controlled by series of neurons firing and synapsing in the brainstem, with little bearing on higher neurological functions. Notably, this means that although the eyes are open, the individual does not need to perceive or respond to visual stimuli to be classified as awake. This objective trait is measurable by behavioral factors (eyes open, motor arousal) and electroencephalography (EEG) findings (exhibiting a signature pattern of wakefulness) (Fernández-Espejo & Owen, 2013).

Being "awake" is not enough. Consciousness requires awareness. Tantamount to awareness is comprehension of self, one's surroundings, and perception of environmental stimuli (Owen, 2013). Contrary to wakefulness, the neurological underpinnings of awareness are a jumble of executive neurological function (certainly requiring activity in the cortex). Currently, the only biomarkers of awareness in clinical practice are behavioral tests such as responding to questions or moving around. These tests are thus effective only for individuals with the capacity for full communication and movement (Dehaene, 2014).

Without wakefulness, one does not have access to sensory input of any kind, and without awareness, he or she cannot process or perceive this input. Therefore, it stands to reason that a irreversible loss of consciousness occurs when either of these components is lost due to death of the brainstem *and/or* essential parts of the cortex. A patient who is determined to be dead by the whole-brain or brainstem criterion would similarly have been seen as dead under this proposed standard, but in addition, so too would a patient with exclusively front-brain death.

DECOUPLING LAW FROM MEDICINE

But if we do not tie the legal standard to a medical one—if we define it conceptually rather than biologically—how exactly would it be applied? This is far from a trivial concern: The meaning and purpose of consciousness has been the subject of centuries of philosophical, ethical, and scientific debate. We are still far from resolving the complex ethical and scientific meaning of consciousness, but remarkable progress has been made recently on identifying a physical marker correlated with the concept (Raffone et al., 2007). Recent studies in cognitive neuroscience have revealed that it may be possible to detect consciousness, even in some individuals that we have previously misidentified as being in a permanent vegetative state. As with other legal standards, a legal standard of death could incorporate medical diagnostic criteria. A legal standard of death could, for example, specify that a legal death occurs when an individual suffers a permanent loss of autonomous respiration and an irreversible loss of consciousness. A medical assessment of an irreversible loss of consciousness is a necessary precondition for pronouncement of legal death. If this were feasible—if science and medicine could offer such a determination—would it be a better way to define the legal end of life?

Dr. Adrian Owen and his research team have been exploring threads of this issue with some troubling results for realizing a present legal standard based on loss of consciousness. Patients are diagnosed as being in a vegetative state by exhibiting a unique continuation of vegetative functions (from electrical activity in the brainstem) together with a complete lack of activation in any part of the cerebral cortex, leaving them without awareness (Fine, 2005). Patients reach a *permanent* vegetative state when they lack executive neural functioning for such an extended time that they are judged unable to regain these functions. Such patients appear to have physical functioning without mental activity; they seem to lack any degree of consciousness that marks a life with present personhood. Using functional magnetic resonance (fMRI) imaging, Dr. Owen and his team sought to determine whether individuals diagnosed as being in a persistent vegetative state in fact lack awareness. The results of his research delivered a shocking blow to modern conceptions of death.

In an article published in *Science* magazine in 2006, Dr. Owen revealed that either some patients diagnosed as being in a persistent vegetative state were misclassified or the idea that those diagnosed as being in a persistent vegetative state lack consciousness may not always hold true. In 2005, Owen chose a patient who was diagnosed as being in a vegetative state after she remained unresponsive with sleep-wake cycles for 5 months and used fMRI scanning to investigate her neural responses as he spoke to her (Owen, 2006). Observing appropriate bilateral activation in speech-specific cortical regions, Owen further prompted her to imagine playing tennis (which activated the supplementary motor area in healthy volunteers) or to imagine navigating her home (which caused volunteers to show activation in the parahippocampal gyrus, posterior parietal cortex, and lateral premotor cortex, indicative of spatial awareness). The subject's brain activity was "indistinguishable" from that of the control volunteers (Owen, 2006), and consistent results across replicated trials indicated that she was clearly aware of researchers' requests. In a replicated study in 2010, four out of twenty-three vegetative state patients mirrored these results, and one successfully replied "yes" or "no" to known questions, substituting one pathway (motor or spatial) for "yes" and the other for "no" (Monti et al., 2010). Using this method for binary communication, Owen scanned Scott Routley, a patient who had been in a vegetative state for more than 12 years, and Routley clearly indicated he was not in any pain (Monti et al., 2010; Walsh, 2012). In 2014, Dr. Steven Laureys elicited distinct behavioral signs of awareness from two vegetative state patients with transcranial direct current stimulation (which encourages neurons to fire using low-level stimulation) on the dorsolateral prefrontal cortex (Thomson, 2014). Although they are hidden, we can surely argue that the awareness levels of these patients constitute life.

These findings suggest that currently, scientists cannot reliably identify patients who have suffered a permanent and irreversible loss of consciousness without extensive brain scanning of each patient. These patients displayed no outward behavioral markers of consciousness and, without scanning or brain stimulation, could be thought to be dead. Nonetheless, at least five individuals scanned by Owen's team had some form of conscious perception and awareness, evidenced by their ability to "answer" questions in the fMRI. Owen has argued that these patients need to be rediagnosed, because their awareness of the world does not fit a diagnosis of vegetative state. This research underscores the limitations of current technology and scientific knowledge in reliably identifying those individuals who have suffered a permanent and irreversible loss of consciousness.

And yet, related studies have echoed Owen's finding of some neurological markers of conscious awareness. For decades, researchers have been fascinated by meditation as a means of purposely altering attention and awareness (Cahn & Polich, 2006). EEG studies have identified wavelengths that are likely linked to consciousness, yet no specific brain regions corresponding to the change of consciousness in meditation had been located, until recently. Observing

Buddhist monks practice mindfulness or concentrative meditation within an fMRI scanner, Dr. Antonio Raffone charted neurological changes as the monks actively cut off their awareness (Raffone et al., 2007). The practice of mindfulness meditation specifically eliminates the sense of self as a separate entity from one's environment. In contrast to meditation that encourages concentration on a specific object (referred to as concentrative), mindfulness meditators remain aware of the sensory and cognitive stimuli around them while simultaneously considering the ongoing flow of thoughts. In this case, "the sense of self shifts from mental thought centered in the body to an impersonal *beingness*" (Cahn & Polich, 2006). These studies noted a distinct deactivation of the medial prefrontal cortex (mPFC) in the case of mindfulness meditation, whereas the mPFC remained active in concentrative meditation and in the control resting state. Thus, the cessation of self-awareness seems to be directly marked by the inactivation of the mPFC. Laureys' study linked the dorsolateral prefrontal cortex to general awareness (Thomson, 2014), accompanied by a P3 wavelength indicative of awareness and measurable with EEG scans (Dehaene, 2014). These are tangible biomarkers for the beginning and ending of awareness that will soon be traceable using EEG versions of Owen's fMRI tests (Fernández-Espejo & Owen, 2013; Goldfine et al., 2013; Pfurtscheller, Brunner, Schlögl, & Lopes da Silva, 2006) or transcranial magnetic stimulation and EEG to observe connectivity between brain regions (Gosseries, 2014). A permanently inactive prefrontal cortex may mark the end of awareness. As a result, it may be an objective and clinical biomarker of the end of life. Such a result could have ethical and legal implications.

Whether or not we are ready to apply these advances to widespread clinical practice, they mark the importance of legal criteria for death that is adaptable to but not defined solely by modern medicine. The respiration standard may suffice for some as a personal mark of death, but it is not legally sufficient because it is constantly challenged and changing with developments in medical technology. A legal standard, therefore, should not be based solely on any medical diagnostic criteria but should instead be defined to serve the purpose for which a legal death is being defined. If we were to legally define death as the *point at which an individual's capacity for consciousness permanently and irreversibly ceases,* we could apply any number of current medical diagnoses to death. If an individual's heart or lungs permanently stop functioning autonomously, he or she will permanently and irreversibly lose consciousness. Similarly, death of either the brainstem or the higher brain will result in loss of wakefulness or awareness, respectively. A legal standard defined to reflect the purposes for which death is being determined would better serve its underlying purpose without being stretched, as medicine continues to stretch the limits of organ revival and survival.

LAW AND THE CIRCLE OF LIFE

How we define the line between legal rights attaching and detaching has significant ethical and legal implications beyond those already mentioned. Consider, for example, a controversial decision by a Texas hospital to refuse the discontinuation of artificial life support for a pregnant woman who had been found unconscious and not breathing for hours. Marlise Muñoz was 14 weeks pregnant when she suffered a pulmonary embolism in November of 2013 and was pronounced brain dead shortly after arriving at the hospital (Fernandez, 2014). Unlike the cases of permanent vegetative state we have examined thus far, Muñoz's diagnosis of brain death indicated permanent cessation of entire brain activity, including the brainstem. Her body required respirators to maintain a set of breathing lungs and a beating heart. Her family fought to follow Muñoz's end-of-life wishes and terminate life support. In a court affidavit, her husband declared, "Over these past two months, nothing about my wife indicates she is alive" (Fernandez, 2014). He argued that the statute was therefore being misinterpreted because one cannot administer life-sustaining treatment to someone who has already died.

Despite her family's requests, the hospital system refused to terminate Muñoz's life support, citing Chapter 166 of the Texas Health and Safety Code, which declares that "a person may not withdraw or withhold life-sustaining treatment . . . from a pregnant patient" (Advance Directives Act, 1999). Muñoz had not written an advance directive, but it would have been no more effective than her family's consent, because the hospital believed that the code overrides patients' advanced directives to protect the lives of fetuses. But if Muñoz were legally dead, would this statute still apply?

In January of 2014, state judge R. H. Wallace, Jr., ruled that Muñoz was legally dead and that the hospital therefore had erred in its reliance on the statute to maintain her life support (Fernandez, 2014). Treatment cannot be "life-sustaining" if an individual has already died.

This case highlights the ambiguities and relevance of the meaning of "legal death" in end-of-life legislation. There is little doubt that Muñoz had suffered an irreversible loss of consciousness, and the more stringent whole-brain death standard was satisfied. But a legal standard of death which was based on irreversible loss of consciousness would have more clearly defined Muñoz as already deceased, rendering life-*sustaining* measures completely irrelevant.

REEXAMINING VIABILITY

But the debate about the end of life is often fought by those who are more concerned about what constitutes the beginning of life. And just as conscious awareness could make it clearer when an individual has reached the legal end of life, it could also be used as a new way to define the legal start of life. The point at which life begins is deeply controversial, particularly as it pertains to the balance of legal rights between a woman and an unborn embryo or fetus. The point at which legal rights attach has implications for how the legal interests of that life are balanced against the legal interests of the mother and therefore has significant implications for the pro-life/pro-choice divide.

Until now, in the United States, the law considered a mother's bodily integrity, privacy, and autonomy interest and the right to terminate pregnancy up until the point at which a fetus is viable and can survive outside the mother's womb. The timing of this *viability standard*, codified by the United States Supreme Court's decision in the case of *Roe v. Wade* (1973), remains in constant flux because of advances in fetal and neonatal medicine. It is a standard that may have had greater moral weight before the advent of invasive technologies that shifted the point of viability to much earlier in a pregnancy. But just as the cardiopulmonary standard of death may have lost its meaning in the face of modern medicine, so too is the viability standard no longer as convenient a bright line to draw. It is little wonder, then, that physicians, legislators, and social advocates continue to search for a more meaningful way to define when the legal rights of a fetus should trump the autonomy, privacy, and bodily integrity rights of the mother.

The most recent suggestion, enforced currently by eleven state laws (Guttmacher Institute, 2015) and a federal bill passed by the U.S. House of Representatives in May 2015, is to define the critical point of fetal autonomy as the point at which a fetus can feel pain. These "Pain-Capable Unborn Child Protection Acts," more commonly dubbed "Fetal Pain Bills," place this point at 20 weeks of gestation, if not earlier. They cite heavy stress hormone releases in fetuses aged 18 to 20 weeks gestation in response to invasive blood transfusions as evidence of fetal pain. (These releases of cortisol and β-endorphins were not detected in fetuses receiving noninvasive blood transfusions and were mitigated when fetuses were treated with analgesics similar to morphine (known μ-opioid receptor agonists) (Fisk et al., 2001). In essence, these laws determine the ability to feel pain to be a sufficient marker for life and, therefore, the beginning of legal rights.

Missing from the current debate is an acknowledgement of the role of conscious awareness necessary to experience pain. As the old philosophical thought experiment goes, if a tree falls in the forest and no one is around to hear it, has it made a sound? This thought experiment raises the important question regarding the roles of observation and perception. In the fetal pain debate, we might ask, if a fetus receives a noxious stimulus but lacks perception, has it experienced pain? Awareness of pain (perception) is a neural experience distinct from processing or experiencing a noxious (unpleasant) stimulus (nociception). Nociception occurs at any point that a noxious stimulus interacts with nociceptors (axon terminals of nerve fibers connecting the skin, joints, muscles, and so on to the spinal cord) (Reddi, Curran, & Stephens, 2013). The nociceptors generate action potentials, which travel up the nerve fibers to the spinal cord, where they synapse onto afferent neurons and are sent up to the brain. Signals from nociceptors are processed in many brain centers. For an individual to perceive pain from a nociceptive action potential, there must be a somatosensory cortex that receives the signal through a functioning thalamus. Only then could the individual perceive the intensity and location of the noxious stimulus (Purves et al., 2008). This thalamocortical connection is essential for the individual to be aware of the noxious stimulus.[2]

And yet, the exact point in fetal development at which the thalamocortical connection is established remains murky. Most literature suggests that sensory nerve fibers and nociceptors have spread across the body by 20 weeks of gestation (Anand, 2004; Bellieni & Buonocore, 2013; Lowery et al., 2007), the thalamus and cortical subplate are complete and sharing neurons by 20 weeks (Bellieni & Buonocore, 2013; Benatar & Benatar, 2001; Kostović & Judaš, 2010), and fibers between the thalamus and cortex are fully developed by 29 to 30 weeks (Kostović & Judaš, 2010).

Although there are still disputes as to the exact timeline of these developments, Fetal Pain Bills fundamentally have pegged fetal pain to nociception rather than perception. But until the signals reach an adequately developed cortex, it is highly unlikely that a fetus perceives the nociceptive signal as a noxious stimulus. And if a fetus cannot perceive pain due to an undeveloped cortex, it cannot consciously perceive *any* stimuli, because pain perception requires the same cortical activation as conscious awareness. If the experience of pain perception is what truly motivates fetal pain advocates, the Fetal Pain Bills should identify perception (or conscious awareness) as the critical moment when fetal interests begin. In other words, a legal life would begin with conscious awareness.[3]

Defining the beginning of a legally autonomous life to be the moment at which the interests of the fetus in life outweigh the privacy interests of the mother would complete the circle of legal life. Legal rights would begin and end with human consciousness.

CONCLUSION

In this chapter, we propose that one could devise a contextual legal standard of death, unhinged from those purely medical ones already in use. Our concern is that legal standards that import wholesale scientific criteria become outdated and disconnected from their original normative meaning through advances in medical discovery and technology. Rather than define the point at which legal rights attach or detach based entirely on medical or scientific criteria, we propose instead that law define death according to normative legal standards. We offer one such approach as a starting point for discussion: a legal standard based on the irreversible loss of the capacity for conscious awareness. We also explore its natural corollary. Just as the end of life has become blurred by changing medical technology, so too has the start of a legal life. The circle of legal life could instead be defined by the beginning and end of the capacity for consciousness.

A legal standard of death would make clear when death has occurred in legal contexts, such as when a murder has transpired, when essential organ donation can begin, when ventilators and other medical interventions can be removed, or when an estate can be distributed. If the legal purposes for which death is defined are understood, a legal standard can be devised to better serve the intended purposes.

A legal standard must also be both administrable and consistent with social norms. If society would not regard an individual as deceased, law cannot sustain a conflicting standard. There are presently scientific hurdles to implementing the standard we propose here. Although there are some early studies on the neural correlates of consciousness, we do not yet have a clear understanding of consciousness, or a clear way to identify its presence or irreversible loss. So we offer this definition to advance the ongoing debate, in the hope that we will provoke more democratic deliberation about the circle of legal life.

NOTES

1. For a detailed discussion, see Bayne and Hohwy's chapter in this volume, "Modes of Consciousness."
2. See also Valerie Hardcastle's chapter in this volume, "Minimally Conscious States and Pain: A Different Approach to Patient Ethics."
3. But see Hardcastle, Chapter 12.

REFERENCES

Ad Hoc Committee. (1968). A definition of irreversible coma: Report of the Ad Hoc Committee of the Harvard Medical School to Examine the Definition of Brain Death. *JAMA*, *205*(6), 337–340.

Advance Directives Act (1999). *Texas Health and Safety Code*. Chapter 166, Sec. 001.166.

Anand, K. S. (2004, January 15). *Expert report of Kanwaljeet S. Anand, M.B.B.S., D.Phil.* National Right to Life Committee. http://www.nrlc.org/uploads/fetalpain/AnandPainReport.pdf

Bass M. (1979). Definition of brain death. *JAMA*, *242*(17), 1850.

Bellieni, C. V., & Buonocore, G. (2012). Is fetal pain a real evidence? *Journal of Maternal-Fetal and Neonatal Medicine*, *25*(8), 1203–1208.

Benatar, D., & Benatar, M. (2001). A pain in the fetus: Toward ending confusion about fetal pain. *Bioethics*, *15*(1), 57–76.

Bernat, J. L., Culver, C. M., & Gert, B. (1981). On the definition and criterion of death. *Annals of Internal Medicine*, *94*, 389–394.

Cahn, B. R., & Polich, J. (2006). Meditation states and traits: EEG, ERP, and neuroimaging studies. *Psychological Bulletin*, *132*(2), 180–211.

DeGrazia, D. (2005). Identity, what we are, and the definition of death. In D. DeGrazia. *Human Identity and Bioethics* (pp. 115–124). Cambridge: Cambridge University Press.

Dehaene, S. (2014). *Consciousness and the brain: Deciphering how the brain codes our thoughts.* New York: Viking—Penguin Group.

Fernandez, M. (2014, January 24). Judge orders hospital to remove pregnant woman from life support. *The New York Times.*

Fernández-Espejo, D., & Owen, A. M. (2013). Detecting awareness after severe brain injury. *Nature Reviews Neuroscience*, *14*(11), 801–809.

Fine, R. L. (2005). From Quinlan to Schiavo: Medical, ethical, and legal issues in severe brain injury. *Baylor University Medical Center Proceedings*, *18*, 303–310.

Fisk, N. M., Gitau, R., Teixeira, J. M., Giannakoulopoulos, X., Cameron, A. D., & Glover, V. A. (2001). Effect of direct fetal opioid analgesia on fetal hormonal and hemodynamic stress response to intrauterine needling. *Anesthesiology*, *95*(4), 828–835.

Goldfine, A. M., Bardin, J. C., Noirhomme, Q., Fins, J. J., Schiff, N. D., & Victor, J. D. (2013). Reanalysis of "Bedside Detection of Awareness in the Vegetative State: A Cohort Study." *The Lancet*, *381*(9863), 289–291.

Goodnough, A. (2005, June 15). Schiavo autopsy says brain, withered, was untreatable. *The New York Times.*

Gosseries O, Thibaut, A., Boly, M., Rosanova, M., Massimini, M., & Laureys, S. (2014). Assessing consciousness in coma and related states using transcranial magnetic stimulation combined with electroencephalography. *Annales Françaises d'Anesthèsie et de Rèanimation*, *33*(2), 65–71.

Guttmacher Institute. (2015). *The most recent suggestion, enforced currently by eleven state laws and a federal bill passed by the U.S. House of Representatives in May 2015, is to define the critical point of fetal autonomy as the point at which a fetus can feel pain.* https://www.congress.gov/114/bills/hr36/BILLS-114hr36pcs.pdf

Kostović, I., & Judaš, M. (2010). The development of the subplate and thalamocortical connections in the human foetal brain. *Acta Paediatrica*, *99*(8), 1119–1127.

Laureys, S. (2005). Science and society: Death, unconsciousness and the brain. *Nature Reviews Neuroscience*, *6*(11), 899–909.

Lowery, C. L., Hardman, M. P., Manning, N., Clancy, B., Whit Hall, R., & Anand, K. J. S. (2007). Neurodevelopmental changes of fetal pain. *Seminars in Perinatology, 31*(5), 275–282.

Merker, B. (2007). Consciousness without a cerebral cortex: A challenge for neuroscience and medicine. *Behavioral and Brain Sciences, 30*(1), 63–134.

Monti, M. M., Vanhaudenhuyse, A., Coleman, M. R., Boly, M., Pickard, J. D., Tshibanda, L., . . . & Laureys, S. (2010). Willful modulation of brain activity in disorders of consciousness. *New England Journal of Medicine, 362*(7), 579–589.

The New Jersey Advance Directives for Health Care and Declaration of Death Acts: Statutes, Commentaries, and Analyses. (1991). Princeton, NJ: New Jersey Commission on Legal and Ethical Problems in the Delivery of Health Care.

New York State Task Force. (1986). *The determination of death.* 2nd ed. New York: New York State Task Force on Life and the Law.

Olick, R. S., Braun, E. A., & Potash, J. (2009). Accommodating religious and moral objections to neurological death. *The Journal of Clinical Ethics, 20*(2), 183–191.

Owen, A. M. (2006). Detecting awareness in the vegetative state. *Science, 313*(5792), 1402.

Owen, A. M. (2013). Detecting consciousness: A unique role for neuroimaging. *Annual Review of Psychology, 64*(1), 109–133.

Pfurtscheller, G., Brunner, C., Schlögl, A., & Lopes da Silva, F. H. (2006). Mu rhythm (de)synchronization and EEG single-trial classification of different motor imagery tasks. *NeuroImage, 31*, 153–159.

Purves, D., Augustine, G. J., Fitzpatrick, D., Hall, W., LaMantia, A., McNamara, J. O., & White, L. E. (Eds.). (2008). *Neuroscience* (4th ed.). Sunderland, MA: Sinauer Associates.

Raffone, A., Manna, A., Perrucci, G. M., Ferretti, A., Del Gratta, C., Belardinelli, M. O., & Romani, G. L. (2007). Neural correlates of mindfulness and concentration in Buddhist monks: A fMRI study. In *Noninvasive Functional Source Imaging of the Brain and Heart and the International Conference on Functional Biomedical Imaging* (pp. 242–244). Joint Meeting of the 6th International Symposium on Source, Hangzhou, China, October 12–14.

Reddi, D., Curran, N., & Stephens, R. (2013). An introduction to pain pathways and mechanisms. *British Journal of Hospital Medicine, 74*(Suppl 12), pC188.

Roe v. Wade. (1973, January 22). United States Supreme Court.

Sisler, J. (2014, March 21). Medical debate over defining death has implications for organ donors. *CBCnews*. CBC/Radio Canada.

St. Petersburg Police Department. (1990, February 25). Incident Report 90-024846. St. Petersburg, FL.Terri Schiavo Life and Hope Network. (n.d.) *Terri's Story—Timeline.* Retrieved from http://www.terrisfight.org/terri-s-story/

Thomson, H. (2014). An electric awakening. *NewScientist, 221*(2958), 8–9. [Originally published online February 26 as "Brain zap rouses people from years of vegetative state."]

Uniform Determination of Death Act. (1980). Uniform Law Commission. The National Conference of Commissioners on Uniform State Laws.

U.S. House of Representatives. (2013). *Pain-Capable Unborn Child Protection Act.* H.R. 1797, 113th Congress.

Veatch, R. M. (1975). The whole-brain-oriented concept of death: An outmoded philosophical formulation. *Journal of Thanatology*, *3*(1), 13–30.

Walsh, F. (2012, November 13). Vegetative patient Scott Routley says "I'm not in pain." *BBC News.*

14

Guardianship and the Injured Brain

Representation and the Rights of Patients and Families

JOSEPH J. FINS AND BARBARA POHL

Severe brain injury may mark a devastating loss of agency for patients, necessitating the appointment of guardians to protect their interests amidst the loss of self-determination. Patients with a disorder of consciousness are often so impaired that they can no longer reliably communicate and make their own healthcare decisions. Their decisional incapacity and the need for reliable and longitudinal decisions about their health and welfare, can prompt the designation of a guardian empowered by the courts to provide direction and guidance about acute and chronic care issues.

Patients with disorders of consciousness share guardianship appointments with other patients who have lost decision-making capacity. But patients with disorders of consciousness do so with a salient difference. Unlike patients with degenerative conditions such as Alzheimer's dementia, patients with severe brain injury do not follow invariably a trajectory of steady decline. Over time, some patients with severe brain injury may improve, evolving from coma to the persistent vegetative state and on into the minimally conscious state and higher levels of functioning (Giacino, Fins, Laureys, & Schiff, 2014). Recent data indicate that increasing numbers of patients with traumatic coma will regain consciousness, with significant proportions regaining functional independence over time (Whyte et al., 2013). Some of these patients may ultimately emerge from a minimally conscious state and recover the ability to functionally communicate, become able to live independently, and regain the capacity to make decisions.

During the time brain-injured patients take to make such a recovery, legal guardians are a prudent way to protect their financial and clinical interests; guardians can advocate for these vulnerable patients and ensure that decision making is careful and prudential. Such safeguards are important because there are a number of errors of omission and commission to which this population is prone and which guardians might prevent or preempt. For example, physicians may misdiagnose patients as permanently vegetative too quickly after injury, thus precluding additional recovery, which may remain possible, and instead recommending a decision to withdraw life-sustaining therapy prematurely. Guardians can also help serve as advocates for patients who might otherwise be relegated to nursing homes or custodial care facilities that do not offer rehabilitation or access to advances in neuroscience that might facilitate their reintegration into civil society (Fins, 2013).

By serving as legal representatives of patients, guardians can find ways to overcome these barriers to care, or what has been viewed as a culture of neglect (Fins & Suppes, 2011). They may, for example, establish and manage trusts that offer physical or speech therapy that facilities would not otherwise provide due to reimbursement policies that eschew the biology of severe brain injury (Fins, 2012). Medical necessity clauses require patients to demonstrate behavioral improvements in time frames that are not appropriate for the recovering brain (Fins, 2015). Guardians may be more capable of questioning insurance companies' denial of care and gaining access to the services facilitating a patient's recovery. Appointing a legal guardian for a brain-injured patient, therefore, may be an important way to "affirm a patient's right to care" amidst a dismissive healthcare system that currently threatens only to promote a patient's "right-to-die" (Fins, 2006).

But all of these protections come with a cost—potential consequences for civil liberties. We assume that in the vast majority of these cases, the motivations behind guardianship appointment are laudatory and beneficent. The pursuit of a guardianship appointment, however, is a significant sociolegal act in which one individual acquires the power to make decisions about an incompetent ward's "person or property" (Garner, 1999). The court, with the power invested in it by the law, takes the rights of democratic citizenship away from individuals who need the help and assistance afforded by guardianship. Because patients with a severe brain injury cannot direct their medical care or manage their finances—at least at some point in time after the injury—these appointments can provide great assistance and even help channel remaining aspects of their ability to be self-determining.

In the context of acquired brain injury, guardianship comes with an additional, possible liability in that some will recover and outgrow the safety and strictures it provides. Because most guardianship appointments are made for patients with degenerative diseases, the premise behind these designations is that the condition for which the guardianship has been secured is a permanent one. If this premise goes unchallenged, then there is the possibility that those patients who could live independently—or begin to assume more responsibility for themselves—will be, under the force of law, permanently divested of their civil rights and liberty through a guardianship appointment.

In this chapter, we argue that guardianship poses four problems for patients living with severe brain injury. First, guardianship coupled with the prevailing nihilism affixed to severe brain injury poses the risk of continuing legal oversight into stages of recovery in which the patient can become more independent. Second, guardians may overrepresent patients who do regain some ability to make certain decisions, and might be able to participate in the decision-making process, thereby denying them their agency. Third, court appointment of (generic) non–family member guardians may undermine family relationships and lead to "best interest" decisions that conflict with the patient's unique preferences known only by family members and other intimates. Finally, institutional requirements for guardianship on admission to chronic care facilities and rehabilitation centers may deny brain-injured patients needed services when they have neither the family nor the resources to secure a guardian. This burden will fall disproportionately on disadvantaged and younger patients, who may not have the fiscal resources or access to legal services necessary for an appointment. Such barriers to care may lead to disparities in outcomes based more on socioeconomic status than on the nature of brain injuries themselves.

BRAIN INJURY AND NIHILISM

Since its origins, the right-to-die movement has been inextricably linked with patients with severe brain injury. One need only recall the landmark case of Karen Ann Quinlan, who was left in a persistent vegetative state (the nomenclature of the day) after a presumed drug overdose. Her parents sought to have the ventilator withdrawn, and the presiding Judge of the New Jersey Supreme Court ordered that Dr. Fred Plum evaluate Ms. Quinlan and determine her diagnosis. Plum was chair of neurology at Cornell University Medical College and co-originator of the vegetative state designation with the Scottish neurosurgeon Bryan Jennett, who was also known for development of the Glasgow Coma scales (Jennett & Plum, 1972; Teasdale & Jennett, 1974). The need for an expert opinion arose over doubts about Ms. Quinlan's diagnosis, its permanence, and its futility before the request of the Quinlan family to remove their daughter from life support was adjudicated (Fins, 2015).

In the Quinlan case, the New Jersey Supreme Court determined that continuation in the vegetative state would be futile and additional recovery impossible. Given this, Judge Hughes ruled in 1976 that it was appropriate to remove the life-sustaining support because the patient could not regain consciousness or return to a "cognitive sapient state" (*In re Quinlan*, 1975–1976). Soon, arguments describing the "irreversibility" of the loss of consciousness and the related "medical futility" of interventions grounded a patient's right to choose when to withhold or withdraw life-sustaining therapies.

Brain injury and the derivative right to die became steeped in a generalized nihilism directed to all patients with severe brain injury (Fins, 2003). Such nihilism propagated the sense of the intractability or immutability of these brain states and of an inevitable downward decline. This led to the notion that decisions about patients with brain injury were invariably end-of-life decisions, when in fact patients may not be at the end of their lives.

Quinlan was also important in defining where the locus of decision making resides and the importance of guardianship for this population. In defining the ways in which the autonomy or self-determination of those who lack decision-making capacity may be respected, right-to-die cases emphasized the significance of determining *who* the courts deemed appropriate to make surrogate decisions. As Marcia Angell (1991, p. 511) acknowledged, "the most important consideration was who made the decision, not what the decision was." In the Quinlan case, for example, the court gave Joseph Quinlan the authority to make the decision to remove, or continue, the ventilator by appointing him as her *guardian* (*In re Quinlan*, 1975–1976). The court's appointment recognized that he would appropriately use substituted judgment and the best interest standard to either withdraw, or continue, care in the absence of his daughter's explicit preferences.

PROGNOSTIC ERRORS

Despite the potential for guardianship to promote and protect the interests of brain-injured patients, it can also undermine these ends by reifying prognostic errors and biases about the permanency of a loss of consciousness. The use of guardianship may problematically affirm the similarity among degenerative conditions and brain injury, leading clinicians to commit an already pervasive prognostic error.

Except in the case of children, the social construction of guardianship carries with it the expectation that an incompetent individual will permanently remain unable to manage his or her medical, financial, or personal affairs. This assumption of permanency is consistent with the demographics of guardianship, in which the majority of guardians *are* designated for individuals with degenerative conditions leading to incompetency. In these cases, the diagnosis is fixed, and the prognosis is one of continued decline. A loss of consciousness for patients with these degenerative diseases often signifies the final turn toward death, triggering discussions about end-of-life care (Posner, Saper, Schiff, & Plum, (2007). Although this is the usual clinical trajectory outside of brain injury, for patients with traumatic brain injury the loss of consciousness can be the harbinger of recovery (as well as decline). Bias occurs when unconscious patients are presumed to have a dire outcome without further exploration of the nature of the disorder of consciousness and its emerging circumstances.

Many clinicians in the acute setting regard unconsciousness after a brain injury as the final harbinger of death rather than a potential beginning of a latent recovery. Similarly, chronic care facilities may carry the false expectation that brain-injured patients have a static diagnosis (Fins, 2009). Unlike many chronic conditions that have fixed diagnoses, brain injury follows a dynamic disease course; patients may begin in a coma and move through different states over time. Currently, clinicians describe these states as aspects of the following disorders of consciousness: coma, persistent and permanent vegetative states, and the MCS (Posner et al., 2007). All of these disorders of consciousness carry different prognoses that may change with the passage of time. Because there is currently no predictive model that can explain the latent recovery of consciousness, clinicians must make prognostic assessments as the injury unfolds (Katz et al., 2009; Whyte et al., 2005).

Guardianships, by their nature as a product of the law, have the virtue of stability, what is described as *stare decisis,* which stresses precedent and stasis. This is necessary for the enduring and efficient practice of guardianship. After all, we can't have shifting guardians with differing decisions all attempting to *stabilize* the shattered world of a patient with brain injury. However, the biology of certain brain injuries is not static but, in fact, mutable, necessitating more flexibility than might exist in the rigid creed of a guardianship designation. A guardianship too grounded in stability may be blind to signs of improvement that herald a change in brain state and have consequences for the self and self-determination (Fins, 2009). These sorts of errors paradoxically can be engendered, in part, by the dynamics of guardianship and the formalism of legal designation. All of this adds a patina of importance to the post, something that is not easily questioned in subsequent moments of more casual observance of evidence of independence.

OVERREPRESENTING PATIENTS

A second problem with guardianship's expectation of permanency is that it may lead to an overrepresenting of the patient at the expense of recognizing a re-emergent self. This is explained in part because one is not incompetent "by degrees" under the law. Although areas of decision-making capacity can be discretely determined in the clinical setting, a determination of competence—with respect to the delegation of authority to a guardian by a court—is global. It is a court's judgment that the individual in all its parts—in toto—does not meet a minimal standard necessary for independent and safe functioning. As a "threshold" concept, competence is not responsive to individuals whose decision-making capability waxes and wanes over time, such as those with bipolar disorder (Wikler, 1979). As Buchanan and Brock (1986, p. 27) noted, individuals cannot possess competence in "matters of degree," although they may possess varying levels of decision-making capabilities.

But whereas a guardian designation takes the person as a whole, clinical care and rehabilitation look at the *recovering parts*, where aspects of the self can regain capabilities that need to be nurtured and sustained for therapeutic and ethical reasons. A patient who is beginning to voice a preference should have these wishes cultivated, interpreted, and understood in a medical model of capacity, in contrast to a juridical model of competence, which is the basis for guardianship determination. Although these "neurodiverse individuals" may make "authentic choices," their deviations from global legal norms can render them wholly incompetent in the eyes of civil society and the law (Banja & Fins, 2013, p. 1376; Mackenzie & Watts, 2011). Guardianship, therefore, does not grant to any incompetent ward—even an individual with some intact aspects of decision-making capacity—a partial stake in the decision-making process.

These nuances are important because patients with severe brain injury may regain decision-making capabilities over time. With rehabilitation, emerging drug therapies, and the use of neuroprosthetic devices, it is even more likely that such individuals will regain consciousness, achieve a form of functional communication, and possess the ability to assent or dissent to procedures. Several interventions, including letter boards, a small cadre of drugs observed to increase awareness, and neuroprosethic devices such as deep brain stimulators, can already facilitate communication with brain-injured patients who possess consciousness, at least at the level of "proof of principle" (Giacino et al., 2012; Monti et al., 2010; Schiff et al., 2007; Whyte & Myers, 2009).

Guardians may over-represent patients' interests by deciding for those who are actually capable. Overrepresenting strips patients of their voice, denying them the opportunity to speak when they are able. This overrepresentation is particularly worrisome because the recovered ability to engage in the decision-making process affords the patient a form of agency that ought to be respected. Because brain-injured patients are liable to experience suffering—what Eric Cassell (1991) described as threats to the self, encouraging their own role in directing care would mitigate some of the associated frustration and isolation.

Because guardians can make decision making more expedient, it may be easy to overlook the occasions when they strip patients of their agency. A patient's ability to assent or dissent raises the possibility that she may disagree with her former self, demanding complex capacity assessments for each choice. Just as guardians need not engage a charge who was deemed incompetent, they could overlook meaningful disparities between past and presently articulated preferences. Prudent surrogate decision makers, by contrast, balance their knowledge of the patient's past wishes with her current preferences and prospects.

Although this mosaic model of decision making is more complex, to entirely neglect a patient's remaining (or regained) voice altogether may only further the patient's isolation. (Fins, 2015) Overlooking the patient's voice fails to recognize how central communication is to being human, as was William Winslade (1998) explained. In sum, the nature of the legal process and the prevailing misconstruals of brain injury compound the risk for guardians to overrepresent brain-injured patients. Guardianship proceedings require great resources and legal representation; it is possible that patients may regain decision-making capacity before a guardian's authority is reversed (Anderson & Fearey, 1989). If guardians endorse misconceptions of severe brain injury, they are even more likely to fail to appreciate the ethical significance of engaging patients in the decision-making process and seeking out their preferences. Guardians may incorrectly regard all brain-injured patients as permanently unconscious, falling prey to the societal neglect syndrome that is built into the history uniting brain injury with the right-to-die movement (Fins, 2003). This issue only gets compounded when a generic guardian—one not naturally associated with the patient—has the decision-making authority.

GENERIC GUARDIANS

A third issue in applying guardianship to brain injury is that the courts may be led to appoint a generic guardian, as opposed to a family member or spouse, for that role. A generic guardian might wrongfully promote the generic best interest of the patient rather than previously expressed wishes or known preferences. Such an appointment may arise because of the aforementioned misconstruals of the biology of severe brain injury. As suggested earlier, one dimension of the legal process establishing a guardian for incompetent patients is selecting an appropriate individual, namely someone who will act in the best interest of the patient. If members of the court misconceive all brain injury as being synonymous with permanent incompetence, however, the selection process may be flawed because their presumption is that consciousness and communication are forever lost. This bias devalues the patient and the patient's past and future relationships with family members.

Because individuals are constructed within families (Nelson, 1994), displacing families from guardianship roles strike us as problematic. Outside the family circle, some choices about "irreversibility" and "futility" may appear to be appropriate. But given the prevailing nihilism, a court might view more "aggressive" care decisions by a surrogate as ones that are inimical to the patient's interest. Because various states and countries have different laws about whether it is appropriate for a court's decision to trump, via guardianship, that of a unified family to make decisions for the patient, it is important to make explicit the biases and inaccuracies that can complicate care decisions about brain injury (Giacino et al., 2014).

When a court bases its decision to appoint a generic guardian on a misinformed conception of brain injury, it may wrongfully overturn a family's unified decision and cause the patient harm. Without intimate knowledge of the patient's past preferences and personality, the court may conclude that something is in the best interest of the patient when it actually runs contrary to what he or she would have chosen. The family's unified decision may accurately reflect what the patient would have wanted. With intimate knowledge of the patient, family members may be best positioned to make challenging healthcare decisions. Many patients with severe brain injury are quite young and have not written an advance directive or designated a healthcare proxy. Even those who have articulated their preferences about end-of-life care may not have anticipated the great prognostic uncertainty of a severe brain injury (Fins, 2013). Thus, family matters.

The court's decision to appoint a generic guardian in such a case problematically denies the patient any representation by those with whom their former selves were close and devalues familial relationships. This devaluation, in our view, stems in part from our sense that the court is devaluing the brain-injured patient. Our speculation about its implicit logic starts with a tautology. The patient is said to permanently lack capacity because he or she is being considered for guardianship. This stems from the perception that loss of consciousness is permanent because of the aforementioned cultural and historical determinants. Therefore, the consequences of not adhering to the patient's prior wishes are low. The patient will never be in a position to be the worse for it, because he or she is forever gone, unconscious and irretrievable. The court may feel free to decide on its own without the moral constraints of countervailing familial wishes *based on knowledge of the patient.*

To some extent, this disregard for the family is reflective of the broader cultural disregard for patients with a disorder of consciousness. But if one views the acknowledgment of consciousness, even at a liminal level, as a civil right, as we do, then to disregard the family and the patient (by proxy) could be construed as a violation of rights and a disrespect of the personhood of those who have been injured (Fins, 2010, 2015).

And it is heart-wrenching, too, beyond the ethical or rights arguments. Imagine how a well-intentioned, dutiful, and loving father or mother might feel being replaced by a court-appointed guardian. We can only imagine the sense of sadness and regret that is superimposed on the primary tragedy that led to the brain injury. Courts should be attentive to the desires of appropriate next of kin to fulfill their responsibility to make important decisions for their loved ones.

Of course, as in cases such as *Schiavo*, familial discord leading to the court's involvement may necessitate appointment of a generic guardian. The court's involvement may reveal a family member who is not acting in a fashion that is in the patient's best interests or consistent with prior wishes. In the Schiavo case, for example, the four *guardians ad litem* attempted to mediate a conflict among family members. They did, however, uphold what her original guardian (Michael Schiavo, her husband), who was familiar with her intimate wishes and preferences, had decided was consistent with what she would have wanted (Gostin, 2005; Wolfson, 2005). Having noted this, it is important that use of guardianship does not allow the courts to undermine the choices of a family member that are consistent with intimate knowledge of the patient when no conflict is present.

DUAL AGENCY

A fourth problem with guardianship is that it may serve as a means for healthcare institutions to protect their own financial interests by requesting guardianship before a patient is admitted to rehabilitation programs. The requirement is intended to ensure that there is someone who is unequivocally fiscally and legally responsible for the patient while under care. This avoids challenges to surrogate consent for treatment and is seemingly more important for assigning responsibility for charges.

Although guardianship stipulations are understandable from an administrative point of view, their presence can exclude disadvantaged or younger patients from receiving care that will facilitate their recovery. These patients are less likely to have the resources to obtain the legal services necessary for guardianship appointments. Such an exclusion would change the demographics of those living with brain injury.

Some institutions, such as inpatient rehabilitation hospitals, already require that admitted patients have guardians to mitigate liability risks. And as noted, with a guardian in place, an institution can hold someone responsible for the cost of the patient's care after insurance benefits cease. The guardian requirement, in this context, may not serve to promote the patient's interest but instead exists to safeguard the institution's own financial well-being (Giacino et al., 2014). This use of guardianship is what is ethically described as dual agency, one in which there can be conflicting allegiances, often at a great burden to families who have to go through a process to secure what should be their "natural right." In this case, the guardian may seem to be in the service of the patient and family, but his or her *raison d'etre* is to serve the institution.

Given that acquiring guardianship is often time-consuming and demands the ability to deal with legal structures, widespread use of guardianship also denies certain patients access to care and skews patient demographics. It potentially discriminates against those who come from families who cannot afford legal fees, do not possess the skills to navigate the legal system itself, and therefore cannot acquire the representation needed for court proceedings. Widespread use of, or requirements for, guardianship would likely further health inequities by barring families of disadvantaged socioeconomic backgrounds from receiving rehabilitative services for their loved ones.

This shift in demographics might initially go undetected given the difficulties of constructing an epidemiology for the various disorders of consciousness, especially the MCS (Fins, Master, Gerber, & Giacino, 2007). Without the ability to effectively assess the numbers of patients living with chronic disorders of consciousness resulting from a severe brain injury, the widespread dissemination of a practice that produces health inequities would be difficult to identify. Because the American healthcare system already contains significant barriers to care, such as medical necessity clauses built into reimbursement schemes (Fins, 2015), it is all the more destructive to introduce other factors denying patients access to care that is instrumental for their recovery.

RECOMMENDATIONS

In light of the problems accompanying the use of guardianship in cases of severe brain injury, we suggest the following recommendations for its proper application. Guardianship is warranted when it is necessary for a family member to protect or provide access to services that the patient would not otherwise receive over the course of his lifespan. This may require the establishment of a trust and associated guardians to oversee the responsible dispersal of funds. In directing care, guardians should not only respect the patient's prior voice but also be responsive to his re-emerging one. This approach is sensitive to the injured brain's potential to recover, encouraging guardians to uncover means to functionally communicate and to continually assess decision-making capacity. When patients are able to communicate their assent or dissent, guardians should weigh their present preferences against those expressed in the past and current safety concerns and best interests. This exercise restores elements of the patient's agency and helps facilitate reengagement with family and community.

When assigning guardianships, courts should give primacy to family relationships and not generic guardians; this respects the specificity of the patient's prior values, commitments, and relationships. Prudent use of guardianship should not conflict with the family's privileged role in the decision-making process.

Any use of guardianship that fails to recognize or cultivate the recovery of consciousness and the related ability of individuals to communicate infringes on the civil rights of brain-injured patients. This misuse of guardianship is inconsistent with the exercise of the law in a pluralistic and inclusive society because it perpetuates societal neglect and segregation of patients with severe brain injury (Fins, 2015). Courts at both the state and federal levels should be self-regulating and should collect quantitative and qualitative data about the use of guardianships for this population to ensure practices consistent with our laws and most cherished democratic values.

ACKNOWLEDGMENTS

Dr. Fins would like to thank the organizers of the Duke Disorders of Consciousness Conference for the opportunity to give the Nancy Weaver Emerson Lecture, Professor Walter Sinnot-Armstrong for his collegiality, and the Buster and Jerold B. Katz Foundations for their support. Dr. Fins and Ms. Pohl also acknowledge partial funding of the Clinical and Translational Science Center (UL1) Cooperative Agreement (CTSC) 1UL1 RR024996 to Weill Cornell Medical College and its Ethics Core.

REFERENCES

Anderson, T. P., & Fearey, M. S. (1989). Legal guardianship in traumatic brain injury: Ethical implications. *Journal of Head Trauma Rehabilitation*, *4*(1), 57–4.

Angell, M. (1991). The case of Helga Wanglie: A new kind of "right to die" case. *New England Journal of Medicine*, *325*(7), 511–512.

Banja, J. D., & Fins, J. J. (2013). Ethics in brain injury medicine. In N. Zasler, D. Katz, & R. Zafone (Eds.). *Brain injury medicine: Principles and practice* (pp. 1372–1390). 2nd ed. New York: Demos Medical Publishing.

Buchanan, A., & Brock, D. W. (1986). Deciding for others. *The Milbank Quarterly*, *64*(S2), 17–94.

Cassell, E. (1991). *The nature of suffering and the goals of medicine*. New York: Oxford University Press.

Fins, J. J. (2003). Constructing an ethical stereotaxy for severe brain injury: Balancing risks, benefits, and access. *Nature Reviews Neuroscience, 4*, 323–327.

Fins, J. J. (2006). Affirming the right to care, preserving the right to die: Disorders of consciousness and neuroethics after Schiavo. *Palliative and Supportive Care, 4*, 169–178.

Fins, J. J. (2009). The ethics of measuring and modulating consciousness: The imperative of minding time. *Progress in Brain Research, 177C*, 371–382.

Fins, J. J. (2010). Minds apart: Severe brain injury, citizenship, and civil rights. In Freeman, M. (Ed.). *Law and neuroscience: Current legal issues 2010*. New York: Oxford University Press.

Fins, J. J. (2012). Wait, wait . . . don't tell me: Tuning into the injured brain. *Archives of Neurology, 69*, 158–160.

Fins, J. J. (2013). Disorders of consciousness and disordered care: Families, caregivers, narratives of necessity. *Archives of Physical Medicine and Rehabilitation, 94*, 1934–1939.

Fins, J. J. (2015). *Rights come to mind: Brain injury, ethics, and the struggle for consciousness*. Cambridge, UK: Cambridge University Press.

Fins, J. J., Master, M. G., Gerber, L. M., & Giacino, J. T. (2007). The minimally conscious state: A diagnosis in search of an epidemiology. *Archives of Neurology, 64*, 1400–1405.

Fins, J. J., & Suppes, A. (2011). Brain injury and the culture of neglect: Musings on an uncertain future. *Social Research, 78*(3), 731–746.

Garner, B. A. (1999). *Black's Law Dictionary*. 7th ed. St. Paul, MN: West Group.

Giacino, J. T., Fins, J. J., Laureys, S., & Schiff, N. D. (2014). Disorders of consciousness after acquired brain injury: The state of the science. *Nature Reviews Neurology, 10*, 99–114.

Giacino, J. T., Whyte, J., Bagiella, E., Kalmar, K., Childs, N., Khademi, A., . . . & Sherer, M. (2012). Placebo-controlled trial of amantadine for severe traumatic brain injury. *New England Journal of Medicine, 366*, 819–826.

Gostin, L. O. (2005). Ethics, the Constitution and the dying process: The case of Theresa Marie Schiavo. *JAMA, 293*(19), 2403–2407.

In re Quinlan. (1975–1976). 348 A.2d 801 (N.J. Super. Ct. Ch. Div. 1975), *modified*, 355 A.2d 647 (N.J.), cert. *denied*, 429 U.S. 922 (1976).

Jennett, B., & Plum, F. (1972). Persistent vegetative state after brain damage: A syndrome in search of a name. *Lancet, 1*(7753), 734–737.

Katz, D., Polyak, M., Coughlan, D., Nichols, M. & Roche, A. (2009). Natural history of recovery from brain injury after prolonged disorders of consciousness: Outcome of patients admitted to inpatient rehabilitation with 1-4 year follow-up. *Progress in Brain Research, 177*, 73–88.

Mackenzie, R., & Watts, J. (2011). Including emotionality in tests of competence: How does neurodiversity affect measures of free will and agency in medical decision making? *AJOB Neuroscience, 2*(3), 1–10.

Monti, M., Vanhaudenhuyse, A., Coleman, M., Boly, M., Pickard, J. D., Tshibanda, L., . . . & Laureys, S. (2010). Willful modulation of brain activity in disorders of consciousness. *New England Journal of Medicine, 362*, 579–589.

Nelson, J. (1994). Families and futility. *Journal of the American Geriatrics Society, 42*(8), 879–882.

Posner, J., Saper, C., Schiff, N. & Plum, F. (2007). Prognosis in coma and related disorders of consciousness, Mechanisms Underlying Outcomes, and Ethical Considerations. In J. Posner, C. Saper, N. Schiff, & F. Plum. *Plum and Posner's Diagnosis of Stupor and Coma* (pp. 341–385). 4th ed. New York: Oxford University Press.

Schiff, N. D., Giacino, J. T., Kalmar, K., Victor, J. D., Baker, K., Gerber, M., . . . & Rezai, A. R. (2007). Behavioral improvements with thalamic stimulation after severe traumatic brain injury. *Nature, 448*, 600–603.

Teasdale, G., & Jennett, B. (1974). Assessment of coma and impaired consciousness: A practical scale. *Lancet, 304*(7872), 81–84.

Whyte, J., Katz, D., Long, D., DiPasquale, M. C., Polansky, M., Kalmar, K., . . . & Eifert, B. (2005). Predictors of outcome in prolonged posttraumatic disorders of consciousness and assessment of medication effects: A multicenter study. *Archives of Physical and Medical Rehabilitation, 86*, 453–462.

Whyte, J., & Myers, R. (2009). Incidence of clinically significant responses to zolpidem among patients with disorders of consciousness: A preliminary placebo controlled trial. *American Journal of Physical and Medical Rehabilitation, 88*, 410–418.

Whyte, J., Nakase-Richardson, R., Hammond, F. M., McNamee, S., Giacino, J. T., Kalmar, K., . . . & Horn, L. J. (2013). Functional outcomes in traumatic disorders of consciousness: 5-Year outcomes from the National Institute on Disability and Rehabilitation Research Traumatic Brain Injury Model Systems. *Archives of Physical Medicine and Rehabilitation, 94*(10), 1855–1860.

Wikler, D. (1979). Paternalism and the mildly retarded. *Philosophy and Public Affairs, 8*(4), 377–392.

Winslade, W. (1998). *Confronting traumatic brain injury.* New Haven, CT: Yale University Press.

Wolfson, J. (2005). Erring on the side of Theresa Schiavo: Reflections of the special guardian *ad litem*. *Hastings Center Report, 35*(3), 16–19.

INDEX